THE
BUTCHER'S
APPRENTICE

TO THE HARD-WORKING FARMERS AND RANCHERS WHO RAISE ANIMALS HUMANELY AND WITH PRIDE AND DIGNITY

JUL 2 6 2016

This edition published in 2015 by
CRESTLINE
an imprint of Book Sales
a division of Quarto Publishing Group USA Inc.
142 West 36th Street, 4th Floor
New York, New York 10018

Printed with permission of and by arrangement with Quarry Books, a member of Quarto Publishing Group USA Inc.

© 2012 by Quarry Books
Text © 2012 Aliza Green

First published in the United States of America in 2012 by
Quarry Books, a member of
Quarto Publishing Group USA Inc.
100 Cummings Center
Suite 406-L
Beverly, Massachusetts 01915-6101
Telephone: (978) 282-9590
Fax: (978) 283-2742
www.quarrybooks.com

ISBN: 978-0-7858-3271-3

Library of Congress Cataloging-in-Publication Data

Green, Aliza.
    The butcher's apprentice : the expert's guide to selecting, preparing, and cooking a world of meat, taught by the masters / Aliza Green ; with photography by Steve Legato.
       pages cm -- (Apprentice)
 Summary: "The masters in The Butcher's Apprentice teach you all the old-world, classic meat-cutting skills you need to prepare fresh cuts at home. Through extensive, diverse profiles and cutting lessons, butchers, food advocates, meat-loving chefs, and more share their expertise. Inside, you'll find hundreds of full-color, detailed step-by-step photographs of cutting beef, pork, poultry, game, goat, organs, and more, as well as tips and techniques on using the whole beast for true nose-to-tail eating. Whether you're a casual cook or a devoted gourmand, you'll learn even more ways to buy, prepare, serve, and savor all types of artisan meat cuts with this skillful guide"-- Provided by publisher.
   ISBN: 978-0-7858-3271-3
 1. Meat cuts. 2. Cooking (Meat) I. Legato, Steve, illustrator. II. Title.
    TX373.G936 2012
    641.6'6--dc23

                                    2011050136

Design: Paul Burgess: Burge Agency
Artwork: Pete Usher
Photography: Steve Legato

Printed in China

THE EXPERT'S GUIDE TO
SELECTING, PREPARING,
AND COOKING A WORLD
OF MEAT

WITH PHOTOGRAPHY
BY STEVE LEGATO

THE BUTCHER'S APPRENTICE ALIZA GREEN

CRESTLINE

CONTENTS

INTERVIEWS WITH THE MASTERS

ILLUSTRATED TECHNIQUES

BEEF

You'll find a broad range of viewpoints in this book, because there is no one right way to eat meat and no one person has all the answers.

INTRODUCTION TO MEAT

In *The Butcher's Apprentice*, I explain the complex nature of producing animals for their meat and how that meat can be prepared so that you can make thoughtful, economical, and nutritionally conscious choices at the market and have the skills to prepare that meat to bring out its best. To write this book, I interviewed breeders, farmers, ranchers, butchers, meat cutters, processors, and other meat industry professionals, and visited as many of their farms and businesses as possible to get a nuanced picture of their lives and work. I asked endless questions to learn more about the challenges they face every day in bringing good meat to market and preparing it properly and safely to enhance its inherent flavor and texture.

The book includes as many meat-cutting techniques as I could find room for. Each technique includes step-by-step instructions illustrated by large, clear photos to give you the confidence to understand the characteristics of various meats, just where on the animal they come from, the nature of individual cuts, and how best to cut, trim, store, and cook the meats, poultry, and game covered here.

You'll "meat" the people behind the product—hardworking, dedicated individuals and companies who care deeply about the animals they work with—to better understand what makes good meat. Everyone that I spoke to was generous in sharing his or her special knowledge and valuable perspective. You'll find a broad range of viewpoints in this book, because there is no one right way to eat meat and no one person has all the answers. The more we know about the meat we eat, the better the choices we can make.

WHAT IS THE MEAT WE EAT?

Meat is mostly composed of bundles of muscle cells called fibers that are made up of two proteins, actin and myosin, which work together to give meat its structure. The filaments are bound together by two types of connective tissue: collagen, or gristle, which is broken down into gelatin by slow cooking, and rubbery elastin, which doesn't break down no matter how long the meat is cooked. Organs such as liver and kidneys, and other parts including blood and—depending on the culture—ears, tail, intestines, tendons, feet, and even eyeballs are also eaten and may be known as offal, viscera, organ meats, or variety meats. Many organ meats are high in cholesterol.

The size and thickness of the muscles determine the texture of the meat, so the meat of young animals, such as lamb or suckling pig, and animals that exercise less, such as domestic turkey and penned rabbits, will be more tender.

Collagen is a long, stiff protein made up of three separate molecules twisted around each other, making for a very stable structure that is difficult to break down. The more collagen contained in a piece of meat, the tougher it will be to cut and chew. Skin and tendons are mostly collagen, and weight-bearing muscles that are in constant use, such as the legs, chest, and rump of cows and pigs, contain the most collagen.

FAT HOLDS FLAVOR

Fat is a concentrated source of energy stored in and around muscles and is key to flavor. When fat is heated, it melts, lubricating muscle fibers and helping to keep the meat moist. Fat content varies greatly depending on species, breed, sex (females are higher in fat than males), method of raising (pasture-grazed and wild animals are lower in fat), the particular cut, trim, and cooking method. Cuts that contain the most fat come from parts that aren't used as much. Ribs are fatty, for example, and outer leg is quite lean.

Grain-finished beef cattle and lamb will be higher in fat while pastured cattle, venison, and bison will be quite lean. Lean meats may have extra fat added by wrapping in sheets of fatback or bacon, inserting small sticks of fat called *lardons* into the meat, slow-cooking in fat as for confit, or by grinding meat with extra fat as for sausage. Fat quantity also depends on how carefully the meat has been trimmed and may be removed after cooking. Chill fatty cuts overnight after cooking and then remove the solidified fat. Pork has been bred to be far leaner than it was in the past, but with leanness comes the tendency to become dry, so the pendulum is swinging back.

Fat dispersed between the muscle fibers is known as marbling. Fat also develops between muscles and underneath the skin (pork fatback). The hard saturated fat surrounding inner organs, especially kidneys (beef and mutton suet and pork leaf lard), is prized for deep-frying and baking because of its high melting and smoke point.

RED, WHITE (AND BLACK) MEAT

Meat is often classified as red or white, depending on the amount of myoglobin, which turns red when exposed to oxygen, is contained in its muscle fiber. Color varies with age, so very young lamb will be pale pink, full-grown lamb will be cherry red, and mature mutton will be dark purplish red. Some pork muscles are darker and close to red in color, like the tenderloin butt, while others are lighter, like pork rib. In general, the older the animal, the darker its flesh and the yellower its fat, though fat color also depends on diet, as young grass-fed beef cattle will also have yellowish fat.

THE COLOR OF RED MEAT

Fresh-cut beef is purplish in color. Oxygen from the air reacts with pigments in the beef and turns it cherry red. The surface color of lamb will be dark cherry-red; pork will be pinkish tan; and veal will be pale pink (for formula-fed veal) to rose colored (for pastured veal). The color of fresh-cut meat is highly unstable and short lived, especially if the meat has been ground, so the interior of a package of ground beef may be grayish brown because oxygen doesn't penetrate below the surface. This is not an indication that the meat is spoiled.

FROM THE FIELD TO THE TABLE AND THE STEPS IN BETWEEN

Meat is produced by killing, or slaughtering, an animal, then separating the edible parts from those to be discarded (which vary from animal to animal and according to local culture and laws). Many cultures have special rituals and laws surrounding the killing of animals and preparation of their meat. In Judaism and Islam, pork is not eaten and animals must be conscious before slaughter, which is done with a sharp knife after saying a prayer. In Hinduism, cows are sacred and are not eaten.

Upon reaching a desired age or weight, livestock are transported to the slaughterhouse, a journey that understandably can be stressful, result in injuries, and have a negative effect on the quality of the meat. Animals from smaller, family-run farms may be slaughtered locally and with more care. The animals are usually stunned, then killed and immediately bled out. Next, the carcass is dressed: usually the head, feet, hide, excess fat, and viscera are removed, leaving muscle and bones.

The next step is butchering: cutting the carcass into smaller, more easily handled parts. Beef, lamb, pork, and veal carcasses are broken down into basic cuts called primals. Beef is divided into eight primals, lamb five, pork four, and veal four. Beef and pork carcasses are split in half lengthwise, and then cut further into wholesale portions. Sheep are cut in quarters across the body so that a rack of lamb includes the ribs on both sides. Depending on the species, the meat is then aged at cold temperatures to increase its tenderness and develop its flavor, as much as six weeks for dry-aged beef.

Meat cutting, or trimming away excess fat, connective tissue, and other unwanted parts, and then preparing it for cooking, is the final step and is covered in detail in this book.

Larger pieces of meat, such as boneless leg of lamb or bottom round of beef, may be tied into a regular shape before cooking. Cuts such as bone-in beef rib (standing rib roast), veal shank, or beef brisket may be cooked whole. A whole lamb or pork shoulder may be slow cooked until tender and much of the fat has melted away. The meat may then be pulled off the bone as for pulled pork barbecue or lamb tacos.

Meat may be sliced through the muscle between the bones as for a bone-in rib chop or it may be cut into small bone-in pieces as for Chinese hacked chicken. Meat may be cut into cubes for kabobs, ground coarsely for chile con carne, medium ground for burgers, or finely ground for pâtés and stuffing. It may be cut into thin slices and pounded for scaloppine. And, meat may be chopped finely and eaten raw for steak tartare and Lebanese lamb kibbe, or thinly sliced and eaten raw for Italian beef carpaccio.

A WORLD OF MEAT EATERS

According to the United Nations Food and Agriculture Organization, the average person in an industrialized nation eats 177 pounds (80 kg) of meat per year, 2.8 times as much as those living in the developing world. Worldwide, pork accounts for 38 percent of meat production, followed by poultry at 30 percent, and beef at 25 percent. Goat meat is a distant fourth but increasingly important, especially in Africa, South Asia, the Caribbean, the Middle East and, now, the United States. The world's supply of about 5 billion hooved and 16 billion winged animals is increasingly being raised in industrial feedlots, which now account for 43 percent of the world's beef and more than 50 percent of its pork and poultry.

Red meat	beef, lamb, goat, pork, and game animals such as venison, pigeon, and duck	Contains more of the narrow muscle fiber that works steadily over long periods without rest as in beef, lamb, and bison—all animals that graze
White meat	chicken, veal, turkey, and rabbit	Contains more broad muscle fiber that works in short fast bursts as in chicken and turkey, animals that have the ability to fly or run quickly for short periods of time
Black meat (*viandes noires*)	French name for the meat of large game animals such as venison, boar, and bear	The meat of these animals is dark purplish in color

Meat is often preserved by curing, smoking, pickling, brining, rubbing with spices, marinating, or grinding with seasonings. It may be cooked and pressed as for Korean beef shank. The head may be cooked and the meat removed, mixed with its naturally jellied juices, and molded for head cheese, brawn, or souse, or the related Ashkenazi Jewish dish, *p'tcha*, or jellied calves feet.

Meat can also be molded or pressed—common for products that include offal such as Scottish haggis (in a pig's stomach) and Pennsylvania Dutch scrapple (made from the bits and pieces of pork)—or canned as for Danish canned ham and French *pâté de foie gras*.

ETHICAL CONSIDERATIONS

There are many ethical and environmental issues surrounding meat eating; some people may choose to eat only ritually slaughtered meat or others may not eat red meat, only chicken and other poultry. Many people eat muscle meats but avoid organs. Some eat meat only from pastured animals, others eat only organic meat, and others eat only meat of heritage breed animals from small, family farms. Some won't eat any blood (in Jewish law, the blood must be removed by salting in order for the meat to be kosher). A new cadre of environmentally aware meat eaters consume meat only that they have hunted and killed themselves.

The UN Food and Agriculture Organization (FAO) estimates that meat production accounts for about 18 percent of the world's total greenhouse gas emissions. Eating lower down on the food chain is beneficial to the environment, and eating lower down on the animal helps to make sure that as much of the whole animal reaches its highest use when prepared in a suitable way. Use the techniques in this book to get the most of the meat you do consume through careful trimming and proper cooking.

HOW TO BECOME A BETTER MEAT EATER

Avoid waste and don't overdo portion size.

Eat less meat and/or choose less expensive cuts (chuck rather than tenderloin, chicken thigh rather than breast) and eat meat from animals that have led the fullest life possible with a minimum of pain and fear.

Buy meat from retailers who carefully monitor the treatment of the animals whose meat they sell or buy directly from the rancher or farmer and consider campaigning to improve animal welfare.

Avoid buying factory-farmed or CAFO (confined animal feeding operation) meats, which degrade the environment because concentration of wastes from these animals impacts air, water, and land quality.

Ask a lot of questions—as a consumer, you have the power to effect positive change in the meats we raise and eat.

Learn the basics of animal anatomy so that you can understand the characteristics of various cuts and the best ways of handling them.

Consider instituting a meatless Monday or other day of the week.

Use smaller amounts of umami-rich meats to flavor other foods, like adding smoked ham hocks or turkey wings to a pot of beans or combining small strips of meat with vegetables and tofu for an Asian stir-fry.

Cheap is fine when it's a lesser cut, such as beef tenderloin tips instead of center-cut tenderloin or lamb shoulder chops instead of rack of lamb, but it's not good when it's about poor conditions, animal stress, environmental degradation, and health and safety issues.

Try a cut you've never heard of to expand your repertoire. You may find something delicious such as beef hanger steak, pork picnic shoulder, or veal cheeks.

Save uncooked and roasted bones and trimmings and poultry carcasses to use immediately or freeze for later use. Simmer slowly until the meat falls off the bone with aromatic vegetables and herbs to make rich, homemade stock, adding body, flavor, and nutrients to soups, sauces, and beans.

HOW TO KNOW IF MEAT IS SPOILED

Meat that is no longer safe to eat will have a strong unpleasant odor and will be sticky or slimy to the touch. A similar smell will be noticeable for a brief time when a vacuum-sealed package of raw meat is first opened, but this is not necessarily a sign of spoilage.

KEEPING THE DANGER OUT OF MEAT

As with all foods, there are certain diseases associated with meat, though with proper handling and cooking, your food will be safe. Escherichia coli, or E. coli, colonizes in the intestines of animals and may contaminate meat at slaughter. E. coli O157:H7 is a rare but virulent strain that can be fatal. Salmonella bacteria may be found in the intestinal tracts of livestock and poultry and inside and on the shells of eggs. Salmonella must be eaten to cause illness, but cross-contamination can occur if uncooked meat (or its juices) mixes with foods such as lettuce that will be eaten raw. E. coli and salmonella can quickly multiply in the "danger zone," between 40°F and 140°F (4.4°C and 60°C), but are destroyed by cooking to higher temperatures.

HOW TO USE SAFE PRACTICES WHEN WORKING WITH MEAT

When buying meat at the supermarket, place packages in disposable plastic bags, if available, to contain possible leaks that could cross-contaminate other foods.

Select meat just before checking out at the supermarket and, if possible, keep it cold on the way home, especially important for ground meat purchased in hot, humid weather.

Take the meat home immediately and refrigerate it at 40°F (4.4°C) or colder. (Commercial meat purveyors store meat at 32°F, or 0°C, to keep it as fresh as possible.)

Before working with meat, wash your hands thoroughly with soap and water. After cutting raw meats, wash hands, cutting board, knife, and countertops with hot, soapy water.

Avoid cross-contamination by keeping raw meats away from other foods and using a separate, washable cutting board (page 10). This is especially important for poultry, which can carry salmonella bacteria.

Avoid eating raw or undercooked meat if you are pregnant or if your immune system is compromised.

HOW TO FREEZE AND THAW MEAT PROPERLY

Cook or freeze poultry, ground meats, and organ meats within two days. Cook beef, veal, lamb, or pork within three to five days.

For best quality with the least loss of juices, freeze meat while it is at its freshest, but wait no more than three days.

Wrap raw meat and poultry securely to maintain quality and to prevent meat juices from getting onto other food, preferably by vacuum sealing (page 15). When freezing meat, vacuum seal or double wrap in special butcher paper, which is designed to prevent leakage, and seal with freezer tape.

To maintain quality when freezing meat and poultry in its original package, rewrap the package again with heavy-duty foil or plastic wrap or place inside a zipper-lock bag and seal.

To thaw frozen meat, thaw in the refrigerator. Ground meat, stew meat, and steaks may defrost within a day. Larger dense cuts, such as brisket, may take two days or longer. Once the meat defrosts, it will be safe in the refrigerator for two to three days before cooking, except for ground meats, which should be cooked within one day of defrosting.

Ideally, cook frozen meat before it has defrosted fully to keep in more of the natural juices. Meat that has been defrosted in the refrigerator may safely be refrozen, though quality may be compromised through loss of juices and changes in texture.

COOKING MEAT TO SAFE TEMPERATURES

Cook beef, pork, lamb and veal steaks, chops, and roasts to at least 145°F (63°C) measured with a food thermometer in the thickest part. For safety and quality, allow meat to rest at least three minutes before carving or eating.

Cook ground beef, pork, lamb, and veal to an internal temperature of 160°F (71°C) measured with a food thermometer in the thickest part.

Cook poultry to an internal temperature of 165°F (74°C) measured with a food thermometer in the thickest part (inside the thigh).

USING SHARP KNIVES AND WELL-DESIGNED, STURDY TOOLS

Professional meat cutters rely on their tools, which don't have to be expensive, though they must be well designed for ease of use, be sturdy, and be reliable. No matter what knife you use, the most important thing is to keep it clean and sharp. The same goes for shears. A sharp knife is a safe knife because it will require minimal effort to cut and will have less chance of slipping.

Use the right tool for the job and respect the knife. Have your knives professionally sharpened to start, then keep them sharp using a sharpening stone or a handheld or electric sharpener. While working, hone the knife often with a metal or diamond sharpening steel or a ceramic sharpening rod to maintain its edge.

A knife made of high-carbon stainless steel will be durable and will retain its edge. The handle should be well shaped to your hand and easy to grip, substantial but not overly heavy. Protect your knives by storing in an individual sheath, in a knife rack, on a magnetic rack, or in a chef's knife roll. If none of these are available, wrap knives individually in a kitchen towel secured with a rubber band.

Work with care, not fear, and always keep your eyes on the knife. Pay close attention to the hand positions shown in the techniques. Notice that the fingers of the nonworking hand are kept curled to prevent injury. When cutting toward your body, use extra care, and, if possible, keep your other hand off the cutting table to prevent possible injury.

1) A boning knife is indispensible for meat cutting and is used for fine trimming and to remove the bones of poultry and meat. It is usually 5 to 6 inches (12.5 to 15 cm) in length with a narrow, curved blade ending in that all-important tip. Boning knives are more or less flexible, depending on the brand and type, which makes it easier to cut around bones and cartilage and through joints. Invest in a top-quality boning knife and it will last through a lifetime of trimming, portioning, and cutting meat.

2) A scimitar (or cimeter) knife gets its name, and original design, from the Arabic term for sword. The long, curved blade, usually 10 to 12 inches (25 to 30 cm) in length, ends in a pointed tip. It is designed to trim and slice larger pieces of meat and poultry and for cutting cold cuts such as ham, turkey, and roast beef into thin, even slices. The wider blade makes it easier to cut through larger pieces of meat such as beef rib eye. While a scimitar is perfectly designed for its tasks, you can get by with substituting a well-designed chef's knife with at least an 8-inch (20 cm) blade.

3) A chef's knife, also known as a French knife or cook's knife, was originally designed to break down large cuts of meat, slice the meat, then cut it into cubes or other small shapes. The blade will usually be between 8 and 12 inches (20 and 30 cm) in length. The two most common blade shapes are French, with a straight edge until the end where it curves up to a pointed tip, and German, with a blade that is continuously curved its entire length. Women, who generally have smaller hands, may find a knife that is lighter in weight and with a shorter blade will be easier to control.

BUTCHER'S STRING. Use special strong natural cotton or linen butcher's string, available at many kitchenware stores. Do not use string made of nylon or polyester, which can leach chemicals into the meat.

The **BUTCHER'S HOOK** shown here has an easy grip handle and a steel hook and is used to help handle larger cuts of meat and to grab on to and pull off heavy membrane.

4) A meat cleaver has a very broad, thick blade, usually rectangular in shape, with a thick edge that is not particularly sharp, so it resists chipping, and is made from softer steel that resists breaking. Its heavy weight is designed to hack through thin or soft bones and connective tissue and to withstand repeated blows. (It is not designed to cut through heavy bones such as leg bones, which are best cut using a saw.) The heavier the cleaver, the easier it is to cut cleanly through bone. Because the cleaver is designed to cut through bone and other body parts, take special care to keep fingers and hands out of the way. A Chinese chef's knife, often known as a Chinese cleaver because of its similar rectangular shape, has a much thinner, lighter blade and is designed for general kitchen cutting, not cleaving through bones and other heavy work.

WOOD OR PLASTIC: THE BEST WORK SURFACE

The best work surface for cutting meat is the traditional close-grained maple cutting board. Dangerous bacteria such as E. coli O157:H7 and salmonella can contaminate any work surface, but scientific studies have shown that bacteria found on a wooden cutting board won't multiply and gradually die. The choice of chefs and meat cutters, an end-grain wood board or block is ideal because the wood fibers, which run in the same direction as the knife cuts, help absorb impact and seal up after cutting, so knives stay sharper longer.

New plastic boards are easily disinfected, but older plastic boards with deep knife scars are difficult to clean and should be replaced. Preferably, use a wooden board or a new or fairly new plastic board that can be run through the dishwasher. If possible, keep a separate board for cutting meat and poultry. To prevent cross-contamination, avoid using the same board to cut raw produce and clean and disinfect the board after each use.

TO CLEAN AND DISINFECT A CUTTING BOARD

Scrape the board with a baker's flat-edged bench scraper or the side of a spatula, then wash the board with a clean sponge soaked in hot water mixed with a little liquid dish detergent. To disinfect, wash or rinse with a mixture of 1 tablespoon (15 ml) bleach and 1 gallon (3.8 L) water. Let stand 2 minutes, then air dry.

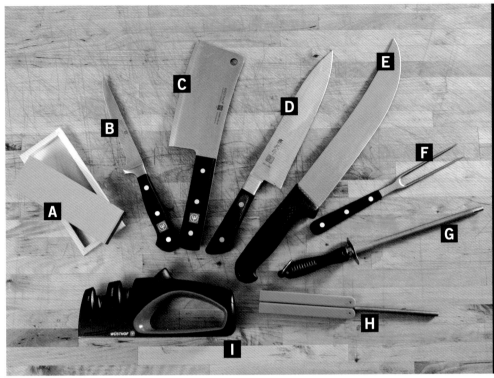

KNIVES AND SHARPENERS

(a) Sharpening stone with fine side showing

(b) Boning knife

(c) Cleaver

(d) Chef's knife

(e) Scimitar knife

(f) Cook's fork

(g) Sharpening steel

(h) Ceramic sharpener

(i) Handheld sharpener

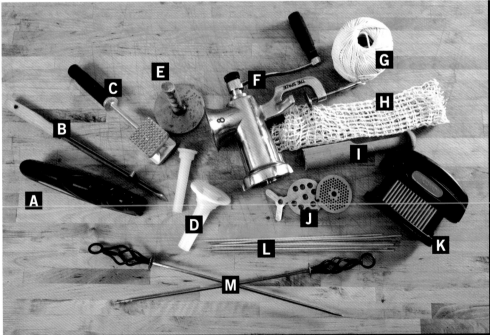

OTHER BUTCHER'S TOOLS

(a) Protective cover for Jaccard cutter

(b) Larding needle

(c) Meat pounder showing knobbed end

(d) Two-part sausage stuffer

(e) Italian brass meat pounder

(f) Italian stainless steel meat grinder

(g) Butcher's string

(h) Net bag for roasts

(i) Wooden pusher for meat grinder

(j) Meat grinder parts: cross-shaped blade, coarse die plate, fine die plate

(k) Jaccard meat tenderizer

(l) Bamboo skewers for satay

(m) Steel skewers for kabobs

MATERIALS NEEDED:

Butcher's cotton string

Scissors

Vacuum sealer and bags, plastic wrap, or zipper-lock bags for storage

1 Using butcher's string and leaving one end attached (the standing end), grasp the free end (the working end) in your dominant hand holding the standing end in your other hand. Run the working end under the center of the meat away from your body.

TYING A BUTCHER'S KNOT

A butcher's knot is used by butchers and chefs to form larger pieces of meat into a compact shape that will cook evenly. The string is always tied against the grain of the meat, so it also acts as a guide as to which direction to slice the meat once cooked. The knot should be secure enough to hold the meat together while it cooks but not so tight that the meat bulges out too much from the sides. Here, we tie a beef tenderloin chateaubriand, usually served for two. Start the first string at the center of the meat, then add strings at either end to secure the meat. On a larger roast, tie more strings in between the center and the ends, keeping them evenly spaced and about 1 inch (2.5 cm) apart. (See also page 80, Preparing a Double Pork Loin Roast.)

5 Insert the working end from the dominant side through the loop that you've created.

6 Pull on the working end while holding the standing end to form a slipknot.

2 Bring the string across the top of the meat and run it behind the standing end on the opposite side of your dominant hand and back to the starting position and twist.

3 Using your nondominant hand, grasp the point where the two strings meet at the twist between your thumb and first two fingers. Hold the string in place up and away from the meat forming an upside-down *V*.

4 Bring the working end of the string down and around the *V* on the side away from your dominant hand, crossing over both strings back to the dominant side of the *V* to form a loop.

7 Secure the string in place using the thumb of your dominant hand placed just in back of the knot toward your body. Grasp the standing end with your dominant hand to yank the knot up tight against the meat.

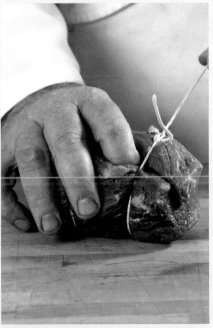

8 To hold the knot in place, make a second loop by bringing the free end up and over the string and through the loop you've created and then pull tight. Cut the string above the knot.

9 Repeat the steps to tie another butcher's knot with a separate piece of string about 1 inch (2.5 cm) away from the first one, continuing until the meat is trussed evenly. Here, we used four lengths of string.

1 Here, we tenderize a dense, somewhat tough but flavorful trimmed beef coulotte, or top sirloin cap (NAMP 184B).

TENDERIZING MEAT USING A JACCARD CUTTER

A Jaccard cutter contains a handle enclosing multiple long, thin, sharp blades (here, forty-five) that move up and down, cutting through dense, tough cuts of meat. Meat that has been "Jaccarded" will be more tender and will absorb marinades thoroughly. The more the meat is Jaccarded, the more tender it will be. However, too much cutting and the meat will turn mushy. Meat that has been Jaccarded will cook much more quickly, so watch carefully to avoid overcooking.

Other cuts that can benefit from Jaccarding are the brisket, sirloin, tri-tip, ranch steak (from the chuck shoulder clod), bottom round, and eye round.

2 Push the Jaccard cutter into the meat against the direction of the grain. Take special care to keep your fingers away from the sharp, spring-loaded cutter blades.

3 Go over the surface several times with the Jaccard cutter. The meat is now ready to portion and cook or marinate and then cook.

USING A VACUUM SEALER TO PRESERVE MEAT

Vacuum sealing is the best way to preserve meat because it removes about 90 percent of the air, preventing oxidation that leads to deterioration and eventually to spoilage, while keeping all the juices intact and the color bright. Here, we vacuum seal double top blade steaks. Ground beef, among the most perishable of meat products, may keep in good condition up to three months if vacuum sealed and kept fully frozen.

To vacuum seal, first pack the product in special precut bags or using a roll of the bag material. Leave several inches (5 to 8 cm) of free space at the head of the bag or roll to make a good seal. Use the smallest bag possible so there's less air to pull out. Keep the side of the bag or roll with the layer of protective netting that helps keep sharp objects such as a bone from poking through the bottom, facing up. Pat the meat to be sealed as dry as possible with paper towels. Very liquidy products are difficult to seal. After sealing, either refrigerate or freeze the meat, always defrosting in the refrigerator on a tray to catch drips.

1 Set up the vacuum machine and open the lid. Place the meat, here two "glued" top blade steaks, into a special vacuum bag or roll just a bit larger than the steaks.

2 Place the bag over the heated sealing bar with the head of the bag inside the shallow trough just beyond the bar.

3 Close the lid and press the start button to seal, supporting the bag from the bottom. The machine will automatically suck out the air from the package and then heat-seal the package.

4 Once sealed, the bag should shrink up tightly around the meat. If the air is not pulled out, the bag may be wrinkled or folded, forming an air pocket. Open the lid and check that the bag is completely flat, then seal again. These vacuum-sealed double top blade steaks are ready for the freezer. Trim off excess bag flap or roll with scissors.

MATERIALS NEEDED:

Transglutaminase powder

Fully trimmed top and bottom top blade muscles

Plastic wrap

Scimitar or chef's knife

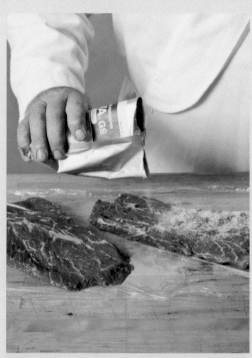

1 Sprinkle 1 to 2 teaspoons (5 to 10 g) of transglutaminase powder evenly over the meat, making sure to sprinkle on the edges.

JOINING TWO PORTIONS OF MEAT
USING TRANSGLUTAMINASE

Transglutaminases are a family of enzymes that bond with proteins. Chefs use this "meat glue" to bind two or more pieces of meat such as sliced bacon wrapped around a filet steak. Unlike gelatin, which also acts as glue, transglutaminase doesn't melt when heated. Here, we "glue" two trimmed top blade muscles, then cut them crosswise to make tender, juicy, and relatively inexpensive double top blade steaks.

Place two trimmed top blade muscles (page 44, Making Flat Iron Steak from Beef Top Blade) on a large piece of plastic wrap with their undersides facing up and the thickest portions facing in opposite directions.

Read the label: Each transglutaminase is formulated for specific uses. Some are meant for fish, some for meat, some for cooked products, some for raw. Some types are sprinkled on the meat to be joined, others are mixed with water and brushed on. Here, we sprinkle the meat with transglutaminase to join two top blade muscles. Purchase transglutaminase from butcher supply companies and online retailers.

5 Wrapped top blade steaks ready to refrigerate. Refrigerate several hours to allow the bond to form.

2 Cover one muscle with the other.

3 Press firmly together to join the two, shaking off and discarding any excess powder.

4 Roll up tightly in the plastic wrap to help join the two muscles together.

6 Unwrap the steak and trim off small slices from both ends to square them off. Cut crosswise into steaks ¾ to 1 inch (2 to 2.5 cm) thick.

7 Double top blade steaks ready to cook.

The savory "meaty" flavor of beef comes from umami, which gets its name from the Japanese word for delicious, umai, and is the fifth flavor after sweet, sour, salty, and bitter.

ABOUT BEEF

Purebred cattle breeds have been selectively bred over many years to emphasize their distinctive identity and the tendency to pass on those traits to their progeny. Smaller British breeds, such as Angus from Scotland and Devon from England, derive from ancient Celtic Shorthorn cattle. Continental breeds such as French Charolais, known since AD 878, and Tuscan Chianina, known for 2,200 years, are large, lean, and muscular, because they were bred as draft animals. The majority of European cattle are local breeds, usually with limited geographical distribution. Wagyu cattle were imported to Japan from Asia in the second century to work the rice fields. America's top five breeds are Black Angus from Scotland, Hereford from England, Limousin from France, Simmental from Switzerland, and Charolais from France.

Beef in North America and the UK comes mostly from 18- to 24-month-old steers (bullocks in the UK) and heifers, weighing about 1,200 pounds (545 kg) and yielding about 770 pounds (350 kg) of meat. About 50 percent are steers, 30 percent heifers, and the remainder dairy cows and bulls. Average weights of Continental cattle are far higher, as much as 3,500 pounds (1,600 kg) for a Tuscan Chianina bull. Beef from older animals, bulls, and dairy cows usually ends up in processed foods. The savory "meaty" flavor of beef comes from *umami*, which gets its name from the Japanese word for delicious, *umai*, and is the fifth flavor after sweet, sour, salty, and bitter. Umami-rich beef has it all: enticing aroma, chewy texture, abundant juiciness, rich mouthfeel, and attractive color.

In North America, the beef carcass is divided at the packing house into eight primal cuts: chuck (shoulder and first five ribs), rib (ribs 6 to 12), loin (rib 13 to the sirloin), sirloin (hip), round (upper leg), brisket (breast), plate (belly), and flank. Cuts from the lower leg, head (see cheeks, page 68), tail, and organs are sold separately. USDA inspection for wholesomeness is mandatory if the meat is to be sold out-of-state; grading is voluntary and is based primarily on marbling (white flecks of fat) within the muscles. Just 2 to 3 percent of American beef is highly marbled Prime, most of which is sold to high-end steakhouses and specialty shops. About 57 percent is moderately marbled Choice, the most common supermarket grade. The remaining 40 percent is lean Select and lower grades, which are ground or processed.

Artisan or whole animal butcher shops (page 70, Tracy Smaciarz) buy beef in quarters, which they break down into smaller cuts at the shop. Their challenge is to find a customer for every cut, because they can't just buy boxes of whatever cut they choose to sell. Much confusion arises because common names for cuts vary greatly so that a Kansas City steak, an ambassador steak, and a New York strip steak are the same. The North American Meat Processors Association has developed standardized specifications for cuts, each with its own number (I have included the numbers in this book).

High-priced beef "middle meats," including the rib, loin, and tenderloin, from the back, are the most tender because these muscles do the least amount of work. Though they represent only about 15 percent of the usable meat on a beef carcass, these cuts sell for about 40 percent of its total value. Cuts from the other primals are denser, tougher, and usually more economical, though price varies, and cuts such as flank, top blade, and short ribs are now in high demand. Tougher cuts contain more collagen (gristle), which breaks down in slow-cooking to provide rich mouthfeel. In general, chuck and rib cuts are fattier, leg and sirloin cuts leaner.

In Continental Europe, beef (and other meats) are seamed to separate the individual muscles. In the UK and North America, meat is often cut with a saw, especially for retail sale, so that many cuts, especially roasts, are made up of multiple muscles, each with its own texture, grain, and direction, which makes them quite challenging to cook and carve properly. European-style seamed cuts, including chuck ranch steak, sirloin tri-tip, and plate skirt steak, are increasingly available. Most steaks and roasts today are sold boneless, cutting down on shipping weight and therefore cost, but also cutting down on flavor and body.

Meat from older, tougher cattle and tough cuts, such as brisket and chuck eye round, call for slow cooking at low temperature using moist heat methods including pot-roasting, braising, and stewing to tenderize the muscles and break down connective tissue. Tender cuts, such as tenderloin and rib eye, are best cooked quickly by dry-heat methods such as stir-frying, grilling, and roasting, but can be dry if overcooked. Medium-tender cuts, such as flank, skirt, and tri-tip, benefit from Jaccarding (page 14) or marinating and may be cooked quickly using dry-heat methods to medium-rare and sliced thinly, always against the grain.

THE EUROPEAN CONVENTION FOR THE PROTECTION OF ANIMALS KEPT FOR FARMING PURPOSES INCLUDES THESE RIGHTS:

Freedom from hunger and thirst—access to fresh water and a diet for full health and vigor

Freedom from discomfort—an appropriate environment with shelter and comfortable rest area

Freedom from pain, injury, and disease—prevention or rapid treatment

Freedom to express normal behavior—adequate space and facilities, company of the animal's own kind

Freedom from fear and distress—conditions and treatment that avoid mental suffering

Photo: Rosalie Winard (ladypelican@gmail.com)

TEMPLE GRANDIN:
EXPERT ON LIVESTOCK BEHAVIOR AND HUMANE SLAUGHTER PIONEER, FORT COLLINS, COLORADO

Dr. Grandin, professor of animal science at Colorado State University, has designed livestock handling facilities worldwide. Close to half of North American cattle are handled in a system that she designed. Her writings on the principles of grazing animal behavior have helped farmers to reduce stress on their animals. She is the author of *Animals in Translation* and *Animals Make Us Human*, which both made the *New York Times* best-seller list. Her life story has been made into an HBO movie titled *Temple Grandin*, starring Claire Danes. In 2010, Grandin was named by *Time* magazine one of its one hundred most influential people.

WHAT IS THE "STAIRWAY TO HEAVEN"?
The stairway to heaven is the curved chute that the cattle walk up in their last moments of life. I design facilities where thousands of animals die. I believe that we are obligated to give an animal a decent life, and the taking of life is a serious thing—it should never be casual. I remember looking over a sea of cattle, thinking that those animals would never have been born if we hadn't bred them.

WHAT DID THE MAJOR FAST-FOOD CHAINS LEARN FROM YOU?
For the CEOs of McDonald's, Burger King, and Wendy's, animal welfare was an abstraction. Then I took them on a visit to the packing houses they used to supply their product. Those executives saw firsthand half-dead dairy cow that couldn't even walk up the ramp, cows that were mooing loudly from fear and pain, and workers who were using wrecked and inadequate equipment. Their eyes were opened because there's nothing like firsthand experience.

DO YOU THINK PEOPLE NEED TO HAVE A CLOSER CONNECTION WITH THE ANIMALS THAT FEED US?
People have associated with domestic animals for thousands of years. Today, most people are far removed from the real world. Nature is very harsh and people who are idealists are often not working in the field. Because everything is abstract, we make unrealistic policies. Kids grow up separated from all things practical and have no idea where anything comes from. I believe in the middle-of-the-road path.

WHY ARE CATTLE FATTENED WITH CORN?
We feed corn to cattle because it's cheap. As the price goes up, they'll be fed much less of it and we'll see more grass-fed cattle. It's that simple. Cattle going to slaughter now are some of the smallest, leanest animals seen in a long time because corn is now being grown for ethanol. We're plowing up pasture land to build ethanol plants. I don't think it's environmentally sound as there's no net energy gain.

WHAT STANDARDS HAVE YOU DEVELOPED FOR MEAT PACKING PLANTS?
In 1996, I was asked to do an audit of meat packing plants for the U.S. Department of Agriculture and concluded that animal welfare evaluations should be judged using a simple, objective scoring system that can be independently audited. I developed a program for the American Meat Institute called Good Management

Practices for Animal Handling and Stunning. We measure three very important things: livestock vocalizations (such as mooing) that may indicate stress, slips and falls that can cause injury, and the accuracy of stunning. At least 95 percent of the cows must be downed with a single shot, no more than three cattle can be mooing in the stun box, and at most one animal can be falling down.

HOW EXPENSIVE IS IT TO ESTABLISH GOOD ANIMAL WELFARE PRACTICES IN MEAT PACKING PLANTS?

You don't have to have elaborate equipment to be good. I've worked on forty-five plants for beef and twenty for pork, making simple changes like installing nonslip flooring on the kill floor and adding lighting that doesn't reflect and frighten the cattle. To avoid fear, we close off the view to the killing floor. We take care of the equipment, making sure everything has a place to be stored properly, that knives are sharpened, and guns are working properly. I took every bit of design ability I had to make these older facilities work without capital improvement. It's amazing what improvements could be made.

CAN THE LARGE FOOD COMPANIES BE A FORCE FOR BETTER ANIMAL WELFARE?

My audit program was first embraced by leading quick-service restaurant chains and later by major retailers, so it's ironic but the big plants were fixed first, and the niche places came along three to four years later. Lots of people think big is bad, but it's badly managed that's bad. Many companies have now installed live-feed cameras so that a third party can make sure that standards are maintained. It's also important for the packing house manager to be involved enough to care, but not so close to day-to-day operations that they become desensitized.

WHAT DO YOU THINK ABOUT ALL THE NICHE MEAT MARKETS? DO THEY MAKE SENSE AND WILL THEY GROW?

Buying local has become a big deal for consumers. The markets for natural, organic, and animal-welfare approved will all grow, and I'm all for it, but we also need to feed low-income people. For that, we've got to have a decent large-scale commercial production. Each segment of the market should sell to its own customers. We've got to stop the bashing. I've worked with them all: One day I'm working with Whole Foods, the next day I'm at McDonald's.

DO YOU EAT AND ENJOY MEAT?

Yes I do. I like a big steak and eat all different meats and have no intention of giving them up. My metabolism needs meat, though I believe some people do better on a meat-free diet.

CAN YOU EXPLAIN TO THE NONPROFESSIONAL READER THE BASICS OF HUMANE ANIMAL SLAUGHTER?

It's about good handling so the animal is calm going in to the kill room. A pig's meat can be ruined in the last five minutes of its life. With fear, they increase the lactic acid in their systems and their meat becomes pale, soft, and watery. Cattle can get totally crazy if they're poked repeatedly with electric prods and their meat toughens. We get rid of distractions and allow them to use their natural tendency to go toward the light. We make sure that the animals are killed with a single shot. It takes care in the actual doing and good management to make sure it's done properly.

HOW DO YOU FEEL ABOUT THE PRACTICE OF SENDING ANIMALS TO HUGE FEEDLOTS?

A lot depends on the location. Feedlots belong in dry places, like Arizona and Colorado. The cattle shouldn't be stomping in the mud. The single most important design feature is that the feed yards should be sloped 3 to 4 percent, because good drainage helps keep cattle healthier.

HOW DO YOU FEEL ABOUT "GRASS-FARMING," THE NAME THAT GRASS-FED BEEF FARMERS USE?

If done right, grazing improves the land. If not, it ruins the land as the cows chew the grass down to dirt. Certain land is just not good for crops and is well suited to grazing. To produce good grass-fed meat that is not tough, stringy, and horrible, we need to use older genetics of breeds like Devon. Those traits are hard to find in modern Angus and Herefords, which have been bred to fatten well on grain and will be too lean if fed on grass. There's always a trade-off. Cattle that are bred to be more hardy and disease resistant will grow more slowly.

WHAT HAVE YOU LEARNED ABOUT ANIMAL BEHAVIOR?

Piglets will be easier to drive if allowed to explore a new floor first. Pigs will be easier to handle if producers walk through finishing pens to teach pigs to quietly move away. Cattle may refuse to walk on grainy floors because of shadows they create. Even yellow tape can frighten cattle because it is unfamiliar. On a sunny day, cattle may refuse to enter a dark plant. Fan blades moved by wind make cattle turn back. Air needs to be directed away from cattle. An animal will look at a sunspot and stop. Noises like those from a pump can scare livestock and cause them to balk. Steers often balk and stop when they see distractions outside fences.

TENDERLOIN OF BEEF

The long lean tenderloin, or filet, which lies on either side of the backbone, is the tenderest muscle in the beef loin section because that muscle does little work. Tenderloin has mild, mellow flavor, very fine grain, and yielding texture. Though high-priced, because it is always in high demand, tenderloin is easy to prepare and there's little waste. There are three main parts of the tenderloin: the head (at the butt end), the center, which is most desirable, and the tail, or tip. Confusingly, the larger "head" end of the tenderloin lies to the rear and the smaller, pointed "tail" or "tip" is in the front, starting just past the rib section. As the tenderloin continues down the length of the back, it becomes larger in diameter and rounder in shape, ending inside the sirloin with a somewhat tougher ball-shaped head end.

The valuable tenderloin is often pulled out of the carcass for sale whole. If left on, the strip and the tenderloin are cut with a saw through the chine (spinal) bones to make the T-bone steak (from the upper end of the strip loin) and the Porterhouse steak (from the lower end of the strip loin). At the sirloin (or butt) end, the tenderloin has a thicker wing section protruding from its side. The best-tasting tenderloin will have the most marbling and will come from Choice or even Prime-graded cattle. If overly lean, tenderloin can be rather bland. A whole roasted tenderloin will shrink much less if cut into individual steaks. If roasting whole, fold the triangular-shaped tail end over itself and tie with a butcher's knot (page 12) to make a roast of even thickness.

Inexpensive, very dark, lean tenderloins may derive from older cows or dairy animals, such as Holsteins rather than more desirable young steers or heifers, and are often served at lower-end steakhouses and found in supermarkets. Tenderloins from European draft breeds such as Chianina and Charolais will be far larger than U.S. tenderloins, which are usually derived from smaller breeds like Angus. The chain, the long thin muscle that runs alongside the main tenderloin muscle, is often included in filet steaks served at less expensive restaurants for better plate coverage. Tenderloin tips cut from the smaller, pointed front end are a good value. Other names include eye fillet (New Zealand), *filetto* (Italian), *filet* (French), *Lungenbraten* or *Filet* (German), and *solomillo* or *filete* (Spanish).

CUTTING TENDERLOIN
INTO INDIVIDUAL PORTIONS

MATERIALS NEEDED:

Scimitar or chef's knife

Vacuum sealer and bags, plastic wrap, or zipper-lock bags for storage

Here, we cut a *chateaubriand*, a steak large enough to serve two, which originally was cut from the sirloin as a French version of British beef steak. Today, in France the chateaubriand is cut from the largest portion of the tenderloin past the head end while the head end is cut into smaller biftecks. We also cut tournedos: smaller, round steaks weighing 2 to 3 ounces (55 to 85 g) from the center toward the tail.

In France, filet mignon was originally cut from the smaller end, known in the United States and the UK as the tenderloin tip or tail. Today, filet mignon may refer to a steak cut from any part of the tenderloin except the tougher head for this French favorite. A large, oval-shaped thinly pounded steak from the tenderloin butt is known as a sabana (blanket) steak in Mexico.

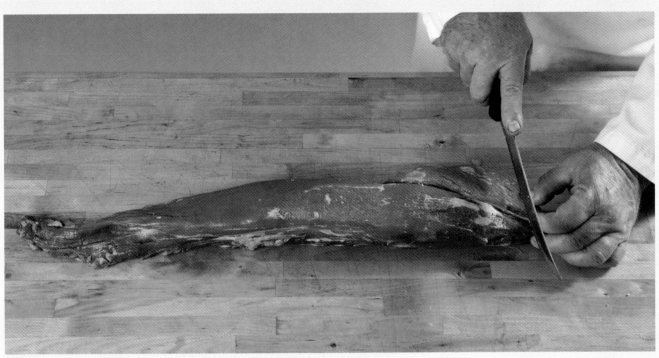

1 Use a scimitar knife (shown here) or a sharp chef's knife to cut a thin slice from the tougher end of the head to square it off and make a more evenly shaped steak.

2 Score the tenderloin before cutting—once cut, a steak can't be uncut. Here, we mark off the chateaubriand, the rounded center-cut, and the thinner tail.

3 Now cut through the tenderloin to make the chateaubriand.

6 After visually dividing the center-cut portion, cut it into four steaks about 2 inches (5 cm) wide, increasing their width toward the narrow tail end for steaks of even weight and cooking time.

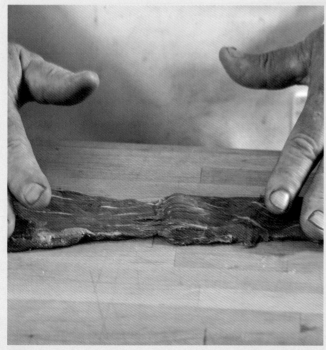

7 Open up the butterflied (horizontally split) tenderloin tip. Flatten slightly and grill or pan-sear.

8 Or, cut into lengthwise strips of even width (here about 0.4 inch, or 1 cm) for stir-fry.

5 Cut two thinner medallions (tournedos in French) from the thicker part of the tail end.

6 Here, we butterfly the tail end. Start at the thicker end of the tail and slice horizontally through the center without cutting all the way through, using your other hand to hold down the top layer.

9 From one tenderloin (right to left), we get one thicker chateaubriand for roasting, four center-cut filet mignon steaks, two thinner medallions, and the tail.

© Rick O'Brien

CARRIE OLIVER:
FOUNDER OF THE ARTISAN BEEF INSTITUTE, SANTA ROSA, CALIFORNIA, AND TORONTO, ONTARIO

Carrie Oliver's background is in new product innovation, but it was her love for a great steak and her passion for breaking new ground that led her to found the Artisan Beef Institute to help people make intelligent buying choices and, as she puts it, "Take the commodity out of beef." Today, many people know only Select, Choice, Prime, and Angus beef. Beef is more like coffee and wine in its complexity—this is something to celebrate.

HOW DID YOU GET STARTED WITH DOING BEEF TASTINGS?

I wanted to buy great-tasting beef raised in a low-stress environment without growth hormones and subtherapeutic antibiotics, but in the supermarket lottery, one week the product was good and the next a disappointment. I went to four or five stores and came home with seven New York strip loin steaks and did a blind tasting with neighbors. None of us agreed on what was best, and we were stunned to find there was no obvious correlation between the grade of beef and flavor or tenderness. The USDA Prime steak was our least favorite.

WHAT IS YOUR GOAL WITH THE ARTISAN BEEF INSTITUTE?

I think of beef as on a continuum. At the top are the best artisanal producers; at the bottom are the least mindful commodity producers. I work to support those in the top 10 to 20 percent. If they succeed, others will follow their lead. Today, a growing number of North Americans are seeking to purchase naturally raised, organic, or grass-fed beef. Yet there are only so many people who will buy beef because of ideology. I want to attract people by their taste buds, sense of entertainment, and desire for community. I want to help people discover that when they know the rancher and butcher behind their beef they cannot only have a better eating experience, they also create a self-reinforcing feedback When the producers know what their customers do or don't like, they can work to improve their beef to drive loyalty, which in turn drives profit.

HOW DO YOU SET UP A TASTING PANEL?

I act as moderator with a panel of experts, ideally a rancher, a butcher, and a chef. Each panelist talks about the top three things they do that influence flavor and texture. We use the same cut of beef from four producers in a blind tasting. We measure texture (is it chewy or tender?), personality (is it reserved or adventurous?), and impression (how fast does it clear the palate?). There is no good or bad rating; we just want people to discover what they like best. We end with the reveal so attendees will know where they can buy their preferred beef.

HOW ARE WINE AND BEEF TASTINGS SIMILAR?

The same things influence wine and beef: grape variety and beef breed, growing region, age, season of harvest, and method and length of aging, to name a few. With beef, low stress leads to better meat. Different wine varieties thrive in some regions better than others. Man developed cattle breeds to flourish in different geographies. We should be celebrating the differences. We don't all want the same kind of coffee or wine. Why should we all look for the same kind of beef?

HOW GREAT IS THE IMPACT OF MARBLING OR GOVERNMENT GRADE ON RELATIVE FLAVOR, TEXTURE, AND JUICINESS?

I discovered via blind tastings that marbling has a far lower impact than I'd been led to believe. After I started saying, "Psst! It's not about the marbling," a friendly meat scientist directed me to a *Journal of Animal Science* study published in 1994, "Effect of marbling degree on beef palatability in *Bos taurus* and *Bos indicus* cattle," which states, "Just 5 to 10 percent of the variation in tenderness can be accounted for by marbling degree." The USDA grading system is based on a visual inspection alone. Would you judge a wine solely on its appearance? If not, then why would we do so with beef?

WHAT ARE SOME OF THE OTHER FACTORS BEYOND MARBLING THAT AFFECT THE QUALITY OF BEEF?

The relative talent of everyone involved in the chain is important, from the people raising the cattle, to the trucker—because poorly handled animals will exhibit flight stress or bruising—to the slaughterhouse professionals, to the butcher, retailer, and cook. A truly talented butcher can look at a carcass and determine how to age the beef and for how long. He or she can also work directly with the customer to decide how to best cut, trim, package, and store the beef.

Once people understand that beef tastes different from farm to farm and that a great rancher and butcher make a big difference, I think we'll see positive changes.

THE MOST IMPORTANT FACTORS FOR GOOD BEEF

Breed: There are dozens, if not hundreds, of cattle breeds and crossbreeds in North America each with its own characteristics and variation between breeds. Tenderness, for instance, is heritable.

Growing region: Shaggy-haired Highland cattle, originally from northern Scotland, might not thrive in Napa Valley, while Brahma from hot-weather India might not be suited to New England.

Diet: The specific grasses, legumes, or grains as well as the minerals in the soil and water

Pharmaceutical use: Growth hormones and other growth stimulants can change marbling, fat content, muscle tone, and impact tenderness.

Age of cattle at slaughter: As cattle age, the muscle fibers and connective tissues change and can affect tenderness and flavor.

Slaughter techniques: Low-stress handling, proper suspension, and cooling of the carcass can significantly affect flavor and texture.

Aging time and technique: Aging helps tenderize beef. Dry-aging can also concentrate the flavors. Most artisan-quality beef will have been aged at least 14 days.

Sex: Most steaks come from steers (castrated males) but a fair number are from heifers (females that have not borne offspring).

MATERIALS NEEDED:

Scimitar or boning knife
Vacuum sealer and bags,
plastic wrap, or zipper-
lock bags for storage

TRIMMING THE TENDERLOIN CHAIN
AND MAKING CUTLETS

The long, thin tenderloin chain that lies
alongside the main tenderloin muscle is
covered with fat. Once removed, the loose
and rather coarse meat can be cut on the
diagonal into small cutlets suitable for
grilling, griddle-cooking, or pan-frying. In
France, the chain, known as the chaînette,
is trimmed, cut on the diagonal, and grilled
as a bistro steak. The chain from a larger
animal may be trimmed, rolled, and tied
into spiral-shaped steaks.

1 Place the tenderloin chain on the work surface with the thicker end
facing your nondominant hand. Slice away the fat covering the chain
using either a scimitar or boning knife, holding the chain in place with your
nondominant hand and keeping your fingers curled for safety.

4 Flattened chain cutlets

2 Slice the chain into sections ⅓ to ½ inch (0.8 to 1.3 cm) thick diagonally cutting in the opposite diagonal as the grain.

3 Arrange the cutlets on the work surface and flatten with the palm of your hand (the chain meat has very loose open grain, so it flattens easily).

5 Tenderloin shown with all usable meat including trim for grinding, chain cutlets, stir-fry strips, tournedos, steaks, and chateaubriand roast.

WHOLE BEEF RIB

The whole beef rib primal (NAMP 103) contains seven bones: numbers 6 to 12. Counting begins at the chuck (shoulder) with numbers 1 through 5. The strip loin starts just beyond number 12 and includes the last rib bone, number 13, which is usually removed. The beef rib is one of two "middle meat" primals along the tender back of the beef cattle along with the short loin or strip loin just past the rib primal.

The front section (ribs 6 to 9) is larger all around, but the valuable central muscle eye is smaller and contains more fat and other muscles. The rear section (ribs 10 to 12) is smaller and leaner, contains a larger eye and less of the surrounding muscles,

and sells for a correspondingly higher price. Because beef rib is more in demand in the winter, summer is a good time to indulge in this prized cut.

Tender, high in fat, and equally high in price, beef rib may be trimmed and roasted whole or cut into individual steaks. A tomahawk steak is the first steak cut from the chuck end and contains the longest rib bone. A cowboy steak is a bone-in rib steak that has been frenched, or cleaned, to the eye, while a barrel steak is a boneless steak cut from the eye at the larger front end. The long bones of the beef rib eye that remain after the eye of meat has been removed are known as dinosaur ribs.

EXTRACTING RIB LIFTER MEAT
FROM WHOLE BEEF RIB EYE

The untrimmed beef rib primal (NAMP 112B) includes the thin rib lifter muscles, which are layered in with the thick fat, or "bark," that covers the rib. Here, we remove the coarse-textured but very flavorful and juicy rib lifter meat, which is excellent cut into cubes for kabobs; butterflied (split horizontally), stuffed, rolled, and braised for Italian beef braciole; grilled and sliced; slow cooked and shredded for barbecue beef; or ground for burgers.

The rounded central eye, seen at the shoulder end, is the most desirable portion of the rib eye. This is a section of the *longissimus dorsi*, the longest muscle in beef cattle and other animals including lamb, pork, beef, veal, goat, and rabbit, which runs from inside the shoulder down the back (dorsi) to inside the sirloin (hip).

The large oblong muscle seen at the loin end is the continuation of the *longissimus dorsi*—just beyond lies the strip loin, or short loin.

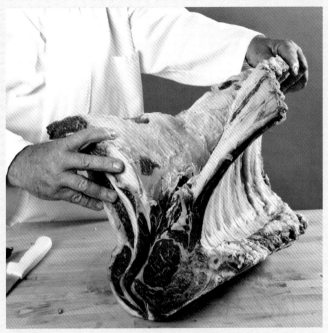

1 Stand the rib with the bone tips facing up. Gripping the knife in your fist for control, cut between the ribs and the fat cap, following the natural seam between the two and using short strokes while pulling away the cap with your other hand.

2 Repeat, sticking the knife farther down between the cap and the rib, then pull apart the top portion of the fat cap from the rib.

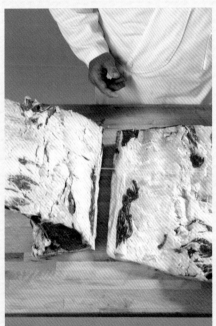

3 Lay the rib down on the worktable, bone-side down, and use the tip of your knife to separate the cap completely from the rib, keeping your knife pointed toward the fat, not the meat.

4 The fat cap with its layers of muscles is on the right, the rib eye muscle with its bones on the left.

5 Lay the fat cap upside down and separate the thin layers of meat from the fat. Begin at the thicker (shoulder) end and insert the tip of the knife just below the top layer of meat.

6 Remove the top layer of muscle and reserve for further trimming.

7 Using the blade of the knife and cutting parallel to the worktable, trim off the fat in sheets, exposing the next layer of muscle.

8 Insert the tip of the knife under the exposed muscle and, keeping the knife parallel to the work surface, cut the muscle away from the underlying fat and then remove.

9 Turn the muscle over and cut the fat away from the other side while gripping the meat with your other hand to hold it in place.

10 The final touch is trimming away the small bits of fat and connective tissue that remain. Repeat, trimming the reserved muscle of its fat and connective tissue.

11 The rib cap with two trimmed larger muscles on the left, the remaining fat cap to the rear, and the smaller bits of trim, suitable for grinding or finely chopping, at the front right.

CUTTING BARREL STEAKS
AND RIB CAP MEAT FROM EXPORT BEEF RIB

The beef export rib (NAMP 109D) has been trimmed of the heavy rib bone ends with their attached muscles (the short ribs, page 51) so the rib weighs less, is less expensive to ship, and is easy to cut into individual steaks. It is usually vacuum-sealed and may be found in warehouse club stores. Because of a buildup of gases, you will notice an odor when you remove the meat from its packaging. This is normal and will dissipate within a few minutes. As for any vacuum-sealed meat, unwrap the beef rib, drain, and blot dry with paper towels before trimming.

The chuck end contains a smaller center eye of muscle covered by a large cap of surrounding muscles; the rib end contains a larger center eye of muscle covered by a smaller cap of surrounding muscles. Here we will be extracting the eye and cutting it into steaks, trimming off the cap meat, and separating out the bone section. The cap meat is ideal for grilling or it may be cut against the grain into strips or cubes for kabobs. The bone section may be roasted, barbecued, or smoked whole or cut into individual ribs.

MATERIALS NEEDED:

Paper towels for blotting

Boning knife

Scimitar, butcher's, or chef's knife

Vacuum sealer and bags, plastic wrap, or zipper-lock bags for storage

1 Rest the rib with the bone side facing up and the chuck end with the smaller muscle eye facing your nondominant hand. Cut between ribs 8 and 9 (the third and fourth ribs counting from the chuck end).

2 Using a scimitar or chef's knife, divide the rib into two sections: the chuck end (at right) with three bones, which will be working with, and the rib end (at left) with four bones, which may be roasted whole or cut into cowboy steaks.

3 Switch to a boning knife and cut between the rib bones and the eye, starting at the loin end. While cutting, pull the main eye section away from the bones until you have separated the two, reserving the bone section for another use.

4 Use a boning knife to cut away the rib cap muscle from the underlying eye, cutting along the natural seam toward the outside to avoid cutting into the eye, which is the most valuable part of the rib.

5 Cut the eye muscle away from the adjoining fat and muscles using short strokes of the boning knife, removing it completely.

8 The trimmed cap meat, shown here, that remains after cutting away the rib eye has moderately loose, coarse texture.

9 The cap meat is ready to cook as is or to be cut into smaller pieces, and is ideal for grilling.

6 Switch to the scimitar or chef's knife (a long blade is ideal for cutting individual steaks in one smooth motion) and, keeping the fingers of your other hand curved for safety, cut individual steaks of desired size.

7 The rib eye barrel has been cut into three steaks shown here, which can be prepared like filet mignon.

10 If desired, cut the cap meat into cubes for stew or kabobs.

11 Thread the cubes onto a large skewer against the grain of the meat, leaving a little space between the cubes so they cook evenly.

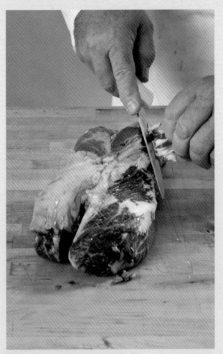

1 Trim off the exterior fat.

TRIMMING THE HANGING TENDER
OR HANGING TENDERLOIN

Also known as hanger steak or by its French name, *onglet*, there is just one small hanging tender weighing not much more than 1 pound (454 g) in each beef carcass. It is technically an organ meat, as it hangs from the kidney and is attached to and supports the diaphragm (or skirt) on the interior of the carcass. Trimmed and cooked, onglet is a French bistro favorite and often kept by butchers for themselves. Until its recent surge in popularity, this cut often ended up in the trim barrel for grinding in North America. Pork hanging tender is also available and may be called the "pillar of the diaphragm." Veal and lamb hanging tenders may be available but will be quite small.

The lean, slightly chewy meat is dark purple with an intense beefy flavor and abundant juices. Its long and prominent grain runs on an angle in a V shape on either side of the tough, elastic membrane that joins its two sides, one larger than the other. The juicy but lean hanger steak is ideal for grilling and pan-searing and best served *à point*, French for medium rare. Because it is an internal organ, the hanging tender spoils easily. Other names include *solomillo de pulmon* (Spain), *lombatello* (Italy), and *entraña gruesa* or thick skirt (Argentina, Uruguay, Paraguay). The hanging tender is also available as a kosher cut.

2 Insert the tip of a boning knife under the silverskin and cut it away in thin strips, using your other hand to secure the meat and keeping your fingers slightly curved for safety.

3 Pull the free end of the silverskin up and away from the meat using your nondominant hand while cutting against the silverskin (not the meat).

4 Using the tip of the boning knife, cut along one side of the tough membrane that joins the two sides, cutting in small nicks against the direction of your other hand. Keep your knife pointed toward the membrane, not the meat.

5 Pull the inner edge of the hanging tender away from the membrane while continuing to cut until you can pull apart the two sections, one cleaned and one containing the membrane.

6 Cut the membrane away from the meat, grasping the freed end in your other hand and pulling away while cutting with your boning knife toward your body. (Stay extra attentive any time you are cutting toward your body.)

7 Hanging tender ready to cook with large section on the left, tough connective membrane in the center to be discarded or perhaps added to the stockpot, and the smaller section on the right. The meat may be cooked whole or you may cut it into medallions, always cutting against the grain, which runs on the diagonal. You may wish to "glue" the two sections together using transglutaminase (page 16).

DRY-AGED BEEF

Beef is dry-aged to tenderize it as it develops bold, concentrated flavor and dense, yet yielding texture. Dry-aging is regulated putrefaction done at carefully controlled temperatures and humidity, usually for two to three weeks. During dry-aging, the meat shrinks substantially as liquid evaporates and develops a hard, dark bacteria-laden crust that must be removed and discarded. About 1 percent of weight is lost for each day of dry-aging, as much as one-third of its weight after trimming. Because only well-marbled beef can be dry-aged, this expensive process is reserved for top-graded carcasses. Whole sides of beef may be dry-aged, though often just the valuable rib and strip loins are used. With the advent of boxed beef in the 1970s, which has been broken down for shipment at the packing house, dry-aging went out of style. Boxed beef is "wet-aged" inside its vacuum-sealed bag for a faster, less costly method of aging without shrinkage. Today, demand for dry-aged beef is on the rise in the search for the utmost in tender, flavorful beef, though its gamier flavor is not to everyone's taste.

Dry-aging is illegal in France, where beef mostly comes from very large breeds such as Charolais that were formerly draft animals. In the UK, the process is known as hanging and is preferred for the finest well-marbled beef. In Australia, some producers now dry-age beef from high-fat, grain-finished animals rather than leaner pastured animals. In the UK, Australia, New Zealand, and occasionally in the United States, mutton is also dry-aged. Game birds such as pheasant may be hung, or dry-aged, usually three to seven days, to increase tenderness and flavor.

CUTTING DRY-AGED BEEF
SHORT LOIN INTO STRIP STEAKS

Here, we cut a dry-aged beef short loin into individual steaks that may be called a strip steak, New York strip steak, sirloin strip steak, ambassador steak, or *tagliata* (Italian).

MATERIALS NEEDED:

Boning knife

Scimitar, slicing, or chef's knife

Vacuum sealer and bags, plastic wrap, or zipper-lock bags for storage

Container for unusable trim

Dry-aged strip loin shell (NAMP 180), or bone-in strip loin, seen from the inner rib end.

This loin has been dry-aged for about two weeks and is marked with a special label seen from the rib end. The meat has a dark red color but not the blackish hard crust that develops after longer aging.

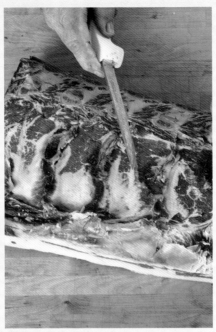

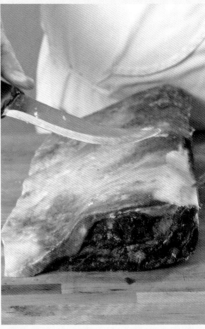

1 Place the loin on its fat side. Trim off excess fat and dark parts. Note: You must discard all trim from dry-aged beef—it CANNOT be used for grinding or any other application.

2 The inner side of the strip loin has been trimmed. Traditionally, several darker crust spots (the knife is pointing at one of them) are left on as a guarantee that the meat was indeed dry-aged.

3 Turn the strip over so its bark side is up and slice off the dark surface fat.

4 Use the scimitar (or a chef's) knife to cut steaks about 1 inch (2.5 cm) thick. Trim each steak following the conformation of the meat, leaving a layer of fat about ¼ inch (6 mm) thick for moisture.

5 Shown here are eleven trimmed steaks cut from one dry-aged strip loin. The more desirable steaks are the first nine counting from the right, with the first steak cut from just past the tender rib and the last from just before the sirloin.

6 The last two steaks from the sirloin end show the beginning of the tougher top butt muscle at the center of the steaks on the fat side. The last steak (on the left) is known as a "veiny" because of the large section of tougher meat it contains.

ABOUT BRISKET

Found in Vietnamese pho soup, Irish corned beef and cabbage, Texas barbecue, Romanian pastrami, and Jewish style sweet and sour sauce, the distinct flavor of beef brisket (NAMP 120) is easily identified no matter how it is cooked and whether served shaved, shredded, or sliced. The brisket contains two dense, coarse-grained muscles: the large, lean oblong first cut, thin cut, or flat (NAMP 120A) and the smaller, fattier, and tougher triangular second cut, navel end, or point (NAMP 120B). The point lies over the front portion of the flat and is connected to it by a thick spongy layer of fat and connective tissue, which helps keep the meat moist when cooked.

The dense, tough, but very tasty brisket must be cooked slowly at low heat and will shrink by about half its original weight, especially after trimming away the surface fat. Start with the largest brisket possible and including both muscles for moisture. Briskets vary greatly in terms of size, conformation, muscle position, and amount of fat. The brisket is equivalent to the front portion of the smaller veal breast. Pork brisket and much smaller lamb brisket are also available. Other names include *Brust* (German), *bianco costata di pancia* or *petto* (Italian), *poitrine* (French), *pecho* (Spanish), thick rib and brisket (UK), and *brustkern* (Austria).

1 Place the boneless whole brisket with its fat-side down and pointed end away from your body. Trim off the heavy layer of fat at the thicker outside edge to access the meeting place of the two muscles.

PREPARING BEEF BRISKET

Here, we prepare boneless beef brisket. Because the grain of the muscles runs in different directions, separate them either before or after cooking. To determine which way to carve—always against the grain—cut a small slice from the front corner of the brisket before cooking to act as a guide.

MATERIALS NEEDED:

Boning knife

Scimitar knife

Vacuum sealer and bags, plastic wrap, or zipper-lock bags for storage

2 Turn the brisket over and cut away the heavy fat at the outside edge. Your goal is to leave a ¼- to ½-inch (0.6 to 1.3 cm) layer of fat for moisture. Cut away any small hard bits that stick out to make a more uniform shape.

3 Turn the brisket over so it rests on its outer fat side. Cut in two horizontally between the two muscles just above the hard chunk of white fat that separates them, exposing a natural seam.

4 Open the brisket, separating the two muscles at their natural seam, cutting with one hand and pulling back with the other.

5 Trim the muscles of excess fat and connective tissue, leaving on some of the fat to provide moisture in cooking.

6 Trimmed brisket point to the left and flat to the right.

GENE GAGLIARDI:
PRESIDENT AND CEO OF CREATIVATORS, COCHRANVILLE, PENNSYLVANIA

Gene Gagliardi is a lifelong meat cutter who took his intimate working knowledge of meat and poultry and created some of new, popular new cuts of meat and poultry. He specializes in working with underutilized, excess raw materials that had minimal value, transforming them into new and highly profitable and often patentable products.

IT SEEMS LIKE MEAT CUTTING IS OFTEN A FAMILY THING WITH KNOWLEDGE PASSING FROM GENERATION TO GENERATION. IS THAT TRUE FOR YOU?

My father, who grew up in southern Italy, had his own slaughterhouse at the age of sixteen. Not much later, he moved to the United States to West Philadelphia, where he opened a corner butcher shop called Gagliardi Brothers. The family lived upstairs and my uncles also worked in the business. At the age of six, I started to work on a pear crate next to my father. He would throw the trimmings on the block I would have to identify what part of animal it came from and cut it up. This gave me my understanding of the various muscles and the best ways of using them.

HOW DID YOU GET YOUR START AS A "MEAT MAGICIAN"?

Eventually, we opened a large wholesale plant supplying McDonald's, Burger King, and Gino's with their hamburgers as the forerunners of portion control products. We were selling over one million pounds [454,000 kg] a week. In 1969, just after building a new 10,000-square-foot [929 square m] plant, we lost several big contracts. With a new plant and no business, I was under a lot of pressure. I came up with the idea for Steak-Umms, made by grinding the meat over and over until it was homogenized, freezing it in a 22-pound [10 kg] log, then slicing it thinly for Philly-style steak sandwiches. My father was convinced that no one would buy it, so I'd work on it at night. The success of Steam-Umms catapulted our sales from $10 million in 1969 to $63 million in 1980.

WHAT ARE SOME OF YOUR OTHER FAMOUS MEAT INNOVATIONS?

I developed the Beef Value Cuts, including the flat iron, the petite tender, and the ranch steak, for the Beef Council using European-style muscle seaming instead of American-style cutting with a band saw. We destroy the carcass when we use the saw. At first, its members responded by saying, "Our butchers can't do this." Now, they all do it. That's why the sign on my wall says, "The impossible takes a little longer."

HOW DO YOU FIND YOUR CLIENTS?

Generally, I come up with ideas and go to them. We do the exploration and creative work that happens before corporate R & D [research and development]. We bring them the seeds and create something totally new.

WHAT OTHER PRODUCTS HAVE YOU DEVELOPED?

For Smithfield Beef, I created the Texas Hold 'Em, a 9-inch [23 cm] grillable short rib, which is scored to the bone, briefly roasted, then grilled, and eats like a steak. That product won the 2008 Research Chefs Association's Big Beef Innovation Contest. Now that the USDA has reduced its recommended pork cooking temperature to 145°F [63°C], I'm working with the Smithfield Packing Company to make new pork burgers and bacon burger toppings. I'm also creating chicken and turkey burgers by finely chopping, rather than grinding, the meat, so it doesn't get all mashed up and maintains texture. When I was told, "There's nothing new you can do with a chicken," I went out and got ten new patents.

WHAT NEW TECHNOLOGY ARE YOU TAKING ADVANTAGE OF?

I'm working with Cuisine Solutions, which specializes in sous-vide products, in which the food is vacuum sealed in airtight plastic. After being slow-cooked at precise temperatures in a water bath, the toughest but most flavorful cuts, such as beef shank, become fork-tender. I remove the heavy cartilage that joins the two sides of a hanger steak and reattach them with Activa [a meat binder developed by the Japanese company Ajinomoto that acts like glue]. I then cut it into 1-inch [2.5 cm]-thick steaks—delicious!

WHAT TRENDS DO YOU SEE IN THE MEAT INDUSTRY?

It seems like our appetite for burgers is endless, so anything different in a burger will be a big seller. Anything in pork that has a bone in it that you can eat like a rib will sell. Bacon continues to be a winner; I don't think it will ever not be. We're also seeing that homemakers now want to get their hands bloody while learning how to cut meat for themselves.

MATERIALS NEEDED:

Scimitar knife

Vacuum sealer and bags, plastic wrap, or zipper-lock bags for storage

MAKING FLAT IRON STEAK
FROM BEEF TOP BLADE

The boneless top portion of a beef chuck is known as a shoulder clod (NAMP 114); the bottom portion is the chuck eye, the continuation of the long main muscle in the rib and short loin sections and its surrounding muscles. The shoulder clod contains three main muscles known as the "Beef Value Cuts," developed for the Beef Council by Gene Gagliardi (page 42): the top blade, the shoulder center, cut into ranch steaks, and the petite, or shoulder, tender. Here, we trim the top blade steak (NAMP 114D), removing the tough white membrane that lies between the two parts of this muscle. At retail markets, the top blade is found cut crosswise into narrow steaks and labeled petite steak but as the membrane is still present, it is less desirable.

1 Place the top blade on the work surface with the fat-side up. Trim the surface fat from the upper side of the top blade.

2 Turn the top blade over and cut away the silverskin, fat, and connective tissue from the underside.

3 Continue trimming away fat and connective tissue from the top blade until fully trimmed as shown.

4 Find the seam toward the back on the thicker end of the top blade and cut away the top layer, leaving the membrane attached to the bottom portion.

5 Pull back the top portion and remove.

6 The two muscles of the top blade. The muscle on the left contains the membrane just below the surface.

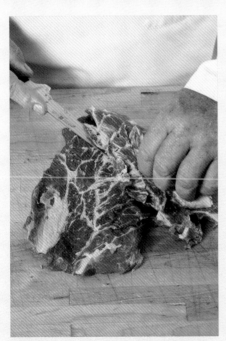

7 Fit the knife under the membrane's edge and cut it away from the underlying muscle, pulling it away with your other hand. Keep your knife parallel to the work surface and pointed slightly up to avoid cutting into the blade meat.

8 Trim both sides of the second top blade muscle of its connective tissue.

9 Two top blade muscles ready to use as is or to be cut into individual portions. After cooking, slice the top blade thinly against the grain, which runs lengthwise.

It's a lot of work, but it's also being needed and accomplishing something positive, so that in the end you get a great feeling of success.

R. L. Freeborn, a fourth-generation rancher from Redmond, Oregon, was the first person to market American Wagyu beef in the U.S. market on a regular basis. Wagyu cattle were first introduced into Japan in the second century from Asia to provide power for the cultivation of rice. Exclusive to Japan for centuries, the breed is world renowned for flavor and tenderness.

WHAT'S A RANCHER'S LIFE LIKE?

It's not a job you can do without losing a lot of sleep. It's a lot of work, but it's also being needed and accomplishing something positive, so that in the end you get a great feeling of success. I have a new dog, but otherwise things are pretty much the same as they've been for years.

DOES KOBE BEEF HAVE RELIGIOUS SIGNIFICANCE IN JAPAN?

Japanese soldiers ate beef to strengthen them for battle. When they returned from war, the soldiers brought along their appetite for beef. Village elders believed that consuming beef indoors desecrated the house and insulted their ancestors. The young men were forced to cook their beef outside on plow shears, which became known as sukiyaki or "plow cooking." The Meiji Restoration in the late 1800s finally relaxed restrictions against eating beef.

R. L. FREEBORN:
OWNER OF KOBE BEEF AMERICA, JAPANESE WAGYU CATTLE RANCHER AND BREEDER, REDMOND, OREGON

WHAT GOT YOU INTERESTED IN RAISING WAGYU CATTLE?

I first tasted Wagyu beef in Japan and found that it was incredibly tender yet also flavourful, and I wanted to bring that beef to [North] America. I bred domestic cattle with Japanese cattle to produce animals with the firmness of domestic beef and the richness and flavor of Wagyu at a competitive price.

WHAT MAKES WAGYU DIFFERENT?

We feed these cattle two to three times longer than commercial cattle with a vegetarian diet and no added hormones or added antibiotics. The beef is wet-aged a minimum of 21 days and grades from 6 to 12 on the Japanese scale [USDA Prime beef is typically 5 plus]. Only about 2 percent of American beef grades USDA Prime; Kobe Beef America's cattle grades upwards of 75 percent Prime. U.S. Prime can be from any number of breeds such as Angus, Hereford, or even Holstein. Kobe Beef America beef is only from the American Wagyu.

WHAT DO YOU LOOK FOR IN BREEDING BEEF?

I think it's important to have a science-based point of view. Most people don't realize how much genetic variation there is among breeds. Each breed has its pluses and minuses; it just depends on the environment, management, and countless other facts. There is an economic reality that we have to consider as breeders: We judge cost by therms [amount of energy produced] to make a pound of beef.

WHAT FACTORS IMPACT THE COLOR OF THE FAT IN BEEF?

Cattle that are entirely grass-fed will have yellowish fat; cattle that are grain-finished will have ivory-white fat from barley and wheat. Cattle fed corn will have a white creamy fat. It really depends on the time of year.

HOW DO YOU FEEL ABOUT USING HORMONES?

We all produce hormones naturally. We [and the cattle] couldn't live without them. Hormones are an important tool in producing commodity beef. We don't use them because they promote growth and body mass; we want marbling and not muscle mass.

CAN YOU TELL ME ABOUT THE HISTORY OF GRAIN-FED BEEF?

Until 1949, 50 percent of American beef was grass-fed; today, 85 to 90 percent is grain-fed. After World War II, income began to rise, people could afford and were willing to spend more for grain-fed beef, and consumers moved away from grass-fed. The flavor profile is different in grass- versus grain-fed beef. It is also the same way for lamb and pork.

WHAT ARE THE MOST IMPORTANT FACTORS IN RAISING BEEF CATTLE?

I illustrate the factors by making a wheel with four main spokes: nutrition, maturity, genetics, environment—animal welfare might be a spoke in between. An Angus top round will be quite different than a Wagyu top round. Just as an Oregon, French, or Australian pinot noir will differ according to where the grapes were grown, so will beef differ according to environment.

HOW IMPORTANT IS IT TO USE THE WHOLE ANIMAL?

I grew up eating tacos made from tripes, tongue, and chorizo—tasty, inexpensive lesser cuts. Now, knowledgeable chefs strive to use the whole animal for philosophical and economic reasons—something that I applaud wholeheartedly. We need to find value items in the rest of the carcass, not just the middle meats. The best buy for the consumer is learning how to use the lesser-known cuts. Most grandmothers know how to use the lesser-known cuts as that is all they had when they were growing up. Remember, ground beef was produced for the poor and low income.

SKIRT STEAK

The long, thin, boneless skirt steak comes from the plate (belly) primal, here beef, though smaller veal, pork, and lamb skirt steak is also available. With its pronounced diagonal grain and a shape that splays out, this cut indeed resembles a wrap-around skirt. A must for Tex-Mex, Korean, and Romanian cooking, it gets its Spanish name, *fajita*, from the word for fan. Skirt, which may be tough if from a very lean animal, takes well to marinades, which are absorbed by the loose, coarse-grained meat. Its strong, beefy flavor works with bold seasonings and its bits of fat keep the meat juicy when grilled. Slice thinly across the grain before serving if desired. Other names include gooseskirt (UK), *lombatello sottile* (Italian), *Gratfleisch* or *Saumfleisch* (German), and *hampe* (French).

1 The thicker outside skirt is on the left with its inner side—the white peritoneum membrane—up. The inside skirt is on the right with its outer side up.

PREPARING BEEF SKIRT STEAK

Here, we prepare outside skirt (NAMP 121C) and inside skirt (NAMP 121D) and cut it in steaks and in finger strips for Korean barbecue. The thicker outside skirt is part of the diaphragm muscle attached to the underside of the ribs. The thick white membrane that covers the meat on both sides may or may not be included. It must be removed if present but once that's done, skirt is one of the easiest meats to trim and cook. The quicker-cooking, thinner inside skirt is a part of the abdominal muscle that extends into the flank.

MATERIALS NEEDED:

Boning knife

Scimitar knife

Vacuum sealer and bags, plastic wrap, or zipper-lock bags for storage

2 To remove the peritoneum, which lies on both the upper and lower sides from the thicker outside skirt, cut a thin slice from the long edge on either side.

3 Cut a small edge of the peritoneum free from the meat at one narrow edge. Hold down the meat with your nondominant hand and pull the membrane back to remove.

4 Turn the skirt over and repeat on the other side, using the boning knife if necessary to cut away any membrane that stays attached.

5 Trim the inside skirt of any membrane and excess fat. Leave some fat to keep the meat moist when cooking.

6 Trim the thicker layers of fat at the edges of the outside skirt.

7 If desired, cut the outside skirt on the diagonal following the grain into individual portions, here four.

8 Trimmed inside skirt. Leave whole, cut in half, or cut into individual portions and wrap for storage.

9 If desired, slice the trimmed inside skirt with the grain into thin finger-like strips for Korean-style barbecue or stir-fry.

BEEF SHORT RIBS

Beef short ribs are cut from the twelve beef ribs and may derive from the meatier but fattier and somewhat tougher chuck short ribs, the leaner back ribs (equivalent to pork baby back ribs), or the plate ribs used here. Retail markets don't usually distinguish which type they are selling. Sharply increased demand among chefs who cook short ribs low and slow often by sous-vide has brought the price up. Short ribs have rich, full-bodied beefy flavor and fall-off-the-bone tenderness when slow cooked. Eastern European Jewish–style flanken (NAMP 1123) are cut across the bones into strips about 1 inch (2.5 cm) wide.

1 Turn a section of plate short ribs over so its inner side is up. Use a small butcher's hook to grab on to the thick white membrane covering the bones. (If not available, use your fingers to grab on to the edge of the membrane and pull.)

PREPARING A ROYAL SHORT RIB
AND BONING BEEF SHORT RIBS

Here we trim a rack of three plate short ribs (NAMP 123), which are squarish in shape with heavy bones and dense, fatty, and collagen-rich flesh.

MATERIALS NEEDED:

Butcher's hook

Boning knife

Butcher's string

Vacuum sealer and bags, plastic wrap, or zipper-lock bags for storage

At George Wells Meats, only one large royal short rib is produced from each three-rib plate, but the remaining two smaller bones can be prepared in a similar fashion.

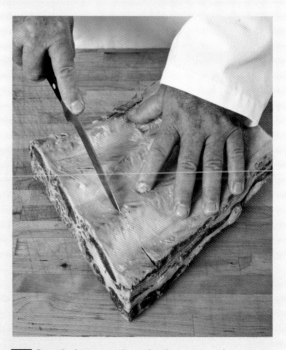

2 Once the heavy membrane has been removed, run your knife across the bones about 1 inch (2.5 cm) from either end, scoring through the remaining membrane to the bone.

3 Use the tip of your boning knife to make a lengthwise score through the membrane that covers each bone.

4 Use the butcher's hook (or your fingers) to pull the membrane from the underlying bone. Repeat with the remaining two bones to make boneless short ribs.

7 Boneless short ribs with crosscut section on the left.

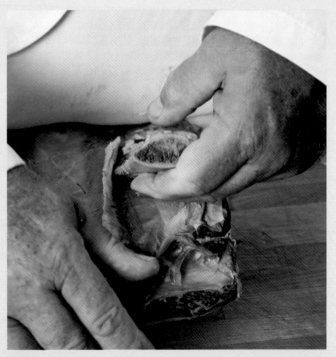

8 To make a royal short rib, pull up the bone from the narrower side at the larger end of the short ribs without pulling it out completely.

5 Slice off the royal short rib.

6 Stick your knife just under the membrane, starting from just under the free end of the bone and slicing to the other end. Remove and discard the membrane.

9 Push the meat down so it partially covers the bone on the inner curved side, then wrap the end over the bone.

10 Tie the meat down to the bone by running the string over the meat and the bone to secure.

GROUND BEEF

Ground beef is made from beef trimmings that are cut into chunks, semifrozen until firm, and put through a meat grinder. For finer texture, the meat is first ground through a coarse die plate, then it is chilled and ground again through a finer die plate. Commercial ground beef may derive from animals from multiple countries of origin and may not be as fresh as is desirable. Because it has so much surface area, ground beef is especially vulnerable to the dangerous strain of E. coli bacteria, O157:H7, that can contaminate the meat at slaughter—another good reason to grind your own. For all ground beef, keep the meat cold (below 40°F, or 4.4°C), use it within two days, or freeze, preferably vacuum sealed. The USDA recommends cooking ground beef to 160°F (71°C) to kill bacteria, though many people prefer their meat cooked to lower temperatures.

Ground chuck has hearty flavor and tends to be higher in fat. Ground sirloin has good flavor but can be quite lean and on the dry side. Ground round is the leanest of all and rather bland and dense because of lack of fat. Whole animal ground beef is made from just that: the entire animal minus only the tenderloin, which is pulled out for separate sale. This type of ground beef is typically produced from pastured cattle on small farms. Many chefs prefer ground meat with 15 to 20 percent fat content, but the only real way to tell how much fat is contained in the meat is to cook it and measure the fat that melts off. If the meat has been ground from the frozen state, the cell walls break down so that the meat leaks juices while cooking.

MATERIALS NEEDED:

Meat grinder, either hand-cranked or electric

Pusher

Coarse and fine grinding plates

Shallow containers lined with parchment, wax, or butcher paper

Vacuum sealer and bags, plastic wrap, or zipper-lock bags for storage

GRINDING BEEF
AND FORMING BURGERS

Here, beef is cut into small pieces, chilled, and then put once through a grinder. The meat must be quite cold, even semifrozen, for grinding; otherwise, the grinder may get clogged. By grinding your own beef or other meat such as lamb, you can control the source of the meat, the cuts used, their freshness, and how much fat is included. Be sure to thoroughly clean the grinder after each use, taking apart all the pieces and washing them in very hot water, especially if grinding pork or poultry. Ground beef may have a small percentage of lactic acid solution added to kill bacteria and maintain its bright, attractive color.

Assemble your chosen meat grinder with a medium plate, clamping it to the table if using a hand-cranked grinder. Because ground beef has so much surface area exposed to oxygen, it will deteriorate quickly, so either vacuum-seal, pack in a zipper-lock bag with all the air squeezed out, or double-wrap with butcher paper and seal with freezer tape for storage in the refrigerator or freezer.

1 Line a tray or other container with parchment, wax, or butcher paper on the work surface. Cut the beef into smaller chunks. If quite lean, add extra fat cubes and combine well. Spread on the tray and semifreeze about an hour or until firm but not hard. (Do not allow the meat to freeze fully as the cell walls will break open and the ground meat won't hold its juices.)

2 Place another paper-lined tray in front of the grinder plate to receive the meat as it is ground.

3 Fill the hopper and use the pusher, if available, to put the meat through the grinder, cranking evenly and steadily. If desired, grind the meat a second time for finer texture. Here, we've left the beef coarse for more bite.

4 If desired, form the ground beef into burgers. Weigh or estimate even portions of meat, here 6 ounces (170 g). Pat lightly but firmly to make hockey-puck-shaped burgers with flat bottoms and tops. Store covered and refrigerated up to 1 day or vacuum-seal and store up to 3 days refrigerated. Or, wrap tightly in plastic and freeze up to 1 month or vacuum-seal and freeze up to 3 months.

Many farmers who graze animals prefer to be known as "grass farmers" because grass is our focus.

Michael Heller's father was a country doctor in Pennsylvania's Lehigh Valley where Heller worked with cattle on a neighbor's farm and loved it. Thirty years ago, Heller was teaching plant ecology at the University of Maryland when the Chesapeake Bay Foundation [CBF] offered him and his wife Clagett Farm to farm and use for teaching sustainable agriculture and helping CBF develop agricultural policy. He now grazes a herd of about seventy-five Red Angus/Red Devon cattle there and is a visiting scholar at Johns Hopkins University teaching sustainable agriculture.

WHY DO YOU CALL YOURSELF A "GRASS FARMER"?

Many farmers who graze animals prefer to be known as "grass farmers" because grass is our focus. Grass and grazing cattle help to build good soils, which in turn help to grow really good grass, which leads to really great cows. Grazing is the least-capital-intensive way to farm. The cows harvest the grass and they naturally fertilize the land with their manure, enriching it and creating soil that is alive with billions of microorganisms. And with a healthy rotational grazing system, parasites are not an issue in the cattle.

HOW DID YOU LEARN TO BECOME A GRASS FARMER?

MICHAEL HELLER:
PASTURED BEEF FARMER, CLAGETT FARM, CHESAPEAKE BAY, MARYLAND

ISN'T IT TRUE THAT GRAIN IS ONLY FED IN THE LAST FEW MONTHS OF THEIR LIVES AND THAT THESE ANIMALS ARE RAISED ON PASTURE UNTIL THEN?

Yes, it's mostly in the last few months that meat animals [cattle and sheep] are fed grain [often in cramped feedlots]. But this grain diet totally changes the environment in their digestive system [ruminants have four-chambered stomachs with the biggest chamber being the rumen]. In a study done by Cornell University, researchers found that digesting corn made the cows' stomachs very acidic so only nasty E. coli bacteria survived. If the cows were fed for just two weeks on only hay without grain, the unhealthy E. coli disappeared. With pasture-based grass-feeding, the cattle have very few health problems.

WHAT IS THE FAT CONTENT OF YOUR CATTLE?

I honestly don't know, as fat is not my focus. My focus is to grow healthy and tasty meat. I do want some marbling in the meat and to get it I must be sure to provide cows high-quality forage throughout their life. However, the single biggest factor in tenderness is not marbling [intramuscular fat] but stress—or lack of it! Animals in feedlots are highly stressed by transport, shots, confinement, and fear, which producers cover up with lots of fat by feeding lots of grain.

WHAT ARE THE DIFFERENCES IN COOKING LEANER GRASS-FED MEAT?

The high fat content in conventionally raised meats requires higher temperatures and longer cooking times to properly cook. With the leaner grass-fed meat it is very easy to overcook. Cook the meat at lower temperatures and for less time, as there is less fat and the meat cooks faster and can lose moisture quickly. For some cuts I like a simple marinade such as olive oil with fresh-squeezed lime juice. You don't want to overdo marinades because you can mask the wonderful natural flavors of grass-fed meats.

PLEASE SUMMARIZE THE GRASS-FEEDING CYCLE

My grazing system is what is called rotational grazing—or management intensive grazing [because it takes real attention, care, and work]—rotating the cows through different pastures every day or two. If they stay longer, cows will degrade the pasture. For good pastured meat, you need to provide rich, high-energy feed consistently throughout the cows' lives and keep the grass at a tender, palatable [for cows] stage.

Our permanent grass pastures have dozens of species of perennials including plants many think of as weeds, such as lamb's-quarter, pigweed, and dandelion, which are all very nutritious when young. I also provide cows with annual millet and cowpea to graze during the hotter and drier summer months when the permanent pastures aren't as lush. In the spring when grass grows abundantly [faster than the cows can eat it], I cut 40 to 50 percent of the young pastures to make dried hay, which supplements the more limited grazing in the winter months.

WHERE DID THE GRASS-FARMING MOVEMENT BEGIN?

Early advocates were the Frenchman Andre Voisin, a researcher and farmer, and Allan Savory, an African wildlife biologist, who both arrived at very similar principles for raising animals by focusing first and foremost on the grass—with the livestock used as a tool to manage the grass. Savory, considered a giant in grazing, used the example of wild herds of bison and antelopes that graze intensely [and densely] in one area and then move on. But really, the grass-based movement has been driven by farmers who are quick to see the benefits to the cows and to the land. And it's the farmers who are experimenting and developing more sophisticated grazing systems. Understanding the complex interactions of plants, animals, and soil requires keen observations and the ability to adapt to continually changing conditions.

MATERIALS NEEDED:

Boning knife

Scimitar knife

Long metal skewers

Vacuum sealer and bags, plastic wrap, or zipper-lock bags for storage

BRAZILIAN SKEWERED PICAÑA
(BEEF COULOTTE STEAK)

The *picaña*, or *punta*, a pointed triangular cap muscle of the top sirloin (NAMP 184D), is a favorite in Brazil and Venezuela for its rich flavor and fine grain. This dense cut is best cooked over natural hardwood charcoal, as is done at *churrascarias* (Brazilian steakhouses). Known as the *coulotte* in English (the same name refers to the nearby tri-tip (NAMP 185C), this cut has the most marbling of the lean top sirloin, so it is flavorful, juicy, and moderately tender. With its dense texture, picaña benefits from marinating and/or from tenderizing with the Jaccard cutter (page 14). The picaña can be trimmed and cooked whole, cut into long, narrow steaks, or sliced, curved, and skewered Brazilian style as in this technique. It may also be cubed for kabobs and stews, coarsely ground or cut into smaller cubes for chili con carne, or sliced into thin strips for stir-fry. If using it for Brazilian-style barbecue, be sure that the picaña includes its fat cap, which helps keep the meat moist when grilling.

1 Place the picaña on the work surface with its fat cap down and its pointed end toward your dominant hand. Begin by cutting the pointed end of the silverskin away from the underlying meat as an aid to removing it.

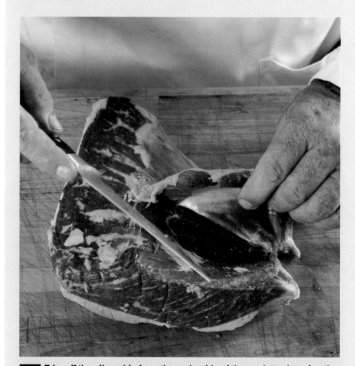

2 Trim off the silverskin from the underside of the coulotte, keeping the knife pointed upward to avoid cutting into the muscle while pulling the thick white silverskin away with your other hand.

3 Turn the picaña over and, using the scimitar knife, trim off the excess fat from the outside in sheets, leaving a layer of fat about ¼ inch (6 mm) thick.

4 Turn the picaña back over and cut into steaks about 1 inch (2.5 cm) thick, always cutting across the grain.

5 Cut about five steaks of even thickness, trimming off the pointed edges of the coulotte to make more evenly shaped steaks. Fold each steak so that its fat is on the outside, curve into a C shape, and skewer, inserting the skewer twice through each steak to secure.

6 Skewered *picaña* ready to grill.

ABOUT VEAL

Veal is the meat of young calves and is appreciated for its mild, adaptable flavor, attractive light color, and collagen-rich texture. It plays a strong role in the cuisines of northern Europe and Italy, which has had a tradition of eating veal since Roman times. Leg of veal is often cut into thin slices (*scaloppine*) that can be quickly sautéed, but veal stew from the shoulder, braised veal shank (*osso buco*), and stuffed, braised veal breast are other popular cooking methods. The chef's basic demi-glace sauce is made from browned veal marrow bones and aromatic vegetables simmered in liquid, which is then strained and reduced until it has a syrupy consistency.

Veal and dairying go hand in hand because to be efficient milk producers, dairy cows must give birth each year. In the past, most heifers (female calves) were kept as milkers, but the males, which are unsuitable for beef production and costly to raise, were sold soon after birth to meat packers as inexpensive "bob-veal," used mostly for processing because of its soft, mushy texture. Today, these calves are raised for veal, usually on smaller family-owned farms. Milk-fed or formula-fed veal comes from these larger, older calves that are fed a special milk supplement. For this type of veal, the lighter the color, the higher the price but the blander the flavor. A calf that grazes and eats solid foods will have darker pink meat with a more pronounced beef-like flavor. It may be called rose, pastured, grass-fed, free-range, natural, or even suckled veal.

In continental Europe, veal often comes from very young animals that are fed on mother's milk. UK rosé veal comes from calves raised on farms associated with the RSPCA's Freedom Food program. In Italy, *vitella di latte*, milk-fed veal, has delicate flavor, tender texture, and pale pink meat with creamy white fat. Reddish *vitellone* comes from older calves and tastes more like beef. Piedmontese *Sanato al Latte* comes from castrated male calves fed milk to keep their meat white, a method at least two hundred years old.

PREPARING RACK OF VEAL

The veal rack, also known as a hotel rack (or rib), may include ribs 5 to 11 for a six-bone rack (NAMP 306B), shown on page 64 from the larger loin end, or ribs 6 to 11 for a six-bone rack (NAMP 306C). The first four or five ribs are included in the chuck or shoulder primal. The equivalent beef rib section includes ribs 6 through 12 (page 30). The loin usually includes two rib bones (numbers 12 and 13). The veal rib eye itself is tender with fine grain and buttery texture. It is surrounded by fattier, tougher, and coarse-grained, though flavorful cap meat, with a greater proportion at the head end.

VEAL AND ANIMAL WELFARE

In the 1950s, American dairy farmers were producing large surpluses of skim milk, which was sold inexpensively to veal producers in the Netherlands. Demand for Dutch milk-fed veal developed in Europe and the method eventually spread to the United States. Farmers moved calves indoors and started confining them in individual crates to save time and space. A strong animal welfare movement grew in the 1980s with the release of photographs of veal calves tethered in crates where they could barely move and sales plummeted. Farmers responded by improving conditions, phasing out or eliminating the use of crates, and raising veal calves without preventative antibiotics or growth hormones. Most veal calves in the U.S. still get a milk-replacement formula rather than mother's milk and may also be fed grain.

Veal calves raised in pens can earn the "certified humane" label from the U.S. organization Humane Farm Animal Care, if farmers follow its strict protocols. In the UK, chefs created the Good Veal campaign arguing that eating humanely raised British veal prevents the calves from being exported to continental Europe or killed shortly after birth.

FRENCHING VEAL RIB BONES

In this technique, we remove the membrane covering the veal ribs bones in a process known as "frenching," which is done for a cleaner, more presentable look.

The exposed bones may be covered with aluminum foil before roasting to keep the edges from burning.

MATERIALS NEEDED:

Boning knife

Sharpening steel

Vacuum sealer and bags, plastic wrap, or zipper-lock bags for storage

1 Starting at the larger (loin) end, cut between the end of the muscle eye and the ends of the bones following the natural seam from one end of the rack to the other.

2 Cut at a rising angle following the line of the eye, not the line of the bones. Following the same cut line, poke the tip of the knife through the membrane between each set of ribs.

3 A shallow slit can be seen when the rack is turned over. Enlarge the slits to make them easier to see.

4 Following the slits like a dashed line, cut straight across through the membrane as far as the bone.

5 Starting at the slit line that you've just created, use the tip of the knife to cut a lengthwise slit along the back of each rib bone, exposing the bone under its cartilage covering.

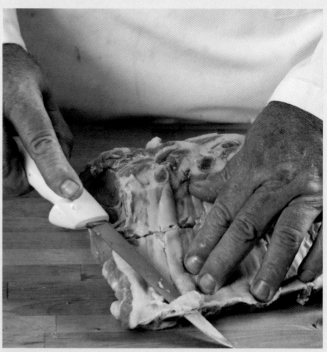

6 Use the tip of the knife to open up the slit at the end of the bone for each rib.

9 Use your knife to detach the bones from any remaining stubborn membrane. The underside of the rib bones toward the chuck end are ridged and may need special trimming.

10 Rack of veal with all bones frenched and membrane still attached.

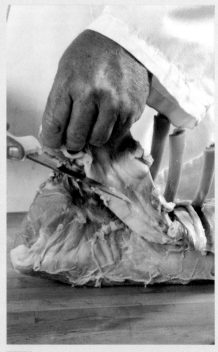

11 Cut away the layer of membrane and connective tissue from the base of the eye.

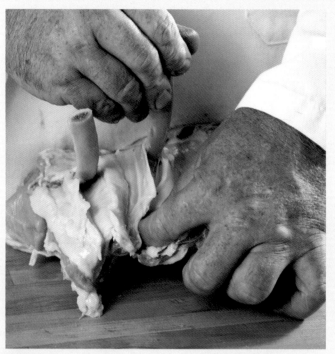

7 Using your dominant hand, one by one, pull the bones up and back, away from its cartilage covering, stopping at the perforated line. If the veal is very fresh, you may need to use a knife to release the cartilage.

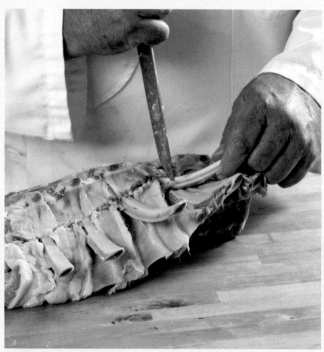

8 Release it from its covering membrane while grasping the bones with your other hand.

12 Cut the membrane between the ribs and the meat, up and around each bone.

13 Rack of veal with bones frenched and trimmed membrane, which is usable for grinding or roasting and added to the pot for demi-glace sauce.

MATERIALS NEEDED:

Boning knife

Sharpening steel

Vacuum sealer and bags, plastic wrap, or zipper-lock bags for storage

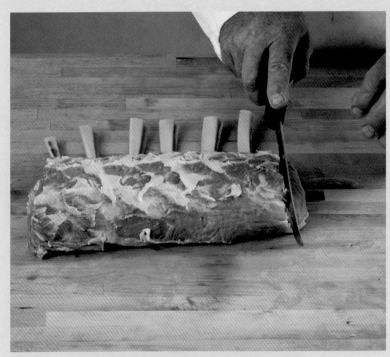

1 For a neater look, cut straight down and remove a thin slice from the larger end of the rack, making a flat end to the eye, which will help make even chops.

CUTTING RACK OF VEAL INTO INDIVIDUAL CHOPS

In this technique, we start with a trimmed and frenched (six-bone) veal rack, NAMP 306B, and cut it into individual chops. The same method works for pork rib chops, goat rib chops, and rack (or rib) of lamb.

4 We will be cutting six chops from this rack. The first four (center-cut) chops shown here are the most valuable, because they have little connective tissue within the eye of the meat. If cutting by weight, you may move the knife to yield chops of the same weight and remove a bone, usually the middle one.

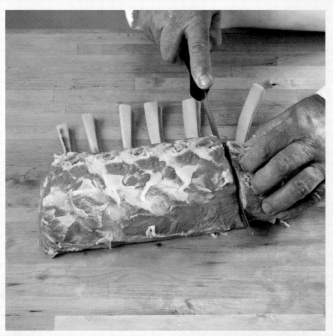

2 Find the place between the vertebrae of the largest and second-largest rib bones at the loin end. Slice down, angling the knife slightly toward the end of the rack facing your nondominant hand.

3 Repeat and cut a second chop, about the same width as the first.

5 Here we cut the shoulder end into two chops. Because these chops contain more connective tissue and are not as desirable, they are ideal for breaded and sautéed veal Milanese.

MATERIALS NEEDED:

Heavy meat cleaver or meat pounder

Platter or tray

Vacuum sealer and bags, ~tic wrap, or zipper-~ock bags for storage

POUNDING VEAL CHOPS

Here, we pound the eye of meat on two rib-end veal chops to flatten and enlarge it. The chops may then be breaded and pan-fried, sautéed, or grilled. Use the same method to pound veal scaloppine from the leg and chicken and turkey cutlets.

1 Place a chop on a wooden work surface with the bone side facing your nondominant hand. A butcher's cleaver works best here, but a meat pounder or the flat side of a heavy chef's knife are alternate choices.

3 Hit the eye with the cleaver several times more to further thin it out, pushing the cleaver out and away while keeping it parallel to the work surface so as not to mash through the meat and make holes.

2 Keeping the cleaver parallel to the work surface, hit the veal eye forcefully and evenly to begin flattening it away from the bone toward the outside edge.

4 Fully pounded veal chops. The larger muscle is the main eye; the surrounding narrow curved muscle is equivalent to the beef rib cap muscle (page 33) and may be prepared the same way.

ABOUT VEAL AND BEEF CHEEKS

Veal (and beef) cheeks are rich morsels of dense, finely grained meat. Look for veal and beef cheeks at whole animal artisan butcher shops (page 70, Tracy Smaciarz), on French bistro menus, in Texas borderlands barbacoa (sometimes made by cooking the entire head and then removing the individual muscles and organs), and in restaurants committed to using the whole animal. Cheeks may be specially ordered and are usually sold frozen with or without their heavy coating of fat removed. They were long one of the cheapest cuts, but due to increased demand, especially from chefs, the price has risen.

Larger and tougher beef cheeks have concentrated "meaty" flavor, while veal cheeks are smaller and more delicate. Other names include masseter muscle (English), *joue* (French), *mejilla* or *cachete* (Spanish), *Rinderbäckchen* (beef cheek, German), and *guancia* (Italian). Italian *guanciale* is pork cheek meat dry-cured like prosciutto.

SOUS-VIDE

Cheeks are braised, barbecued, or cooked using other long, slow cooking methods such as *sous-vide*. This French term means "under vacuum" and refers to a method of cooking food sealed in airtight heavy plastic bags very slowly in a water bath at carefully controlled low temperatures to help dissolve the connective tissue and tenderize tough but flavorful cuts of meat. Like all braised meats, cheeks reheat well, and are often even better the second time around, and freeze well.

TRIMMING VEAL (OR BEEF) CHEEKS

Here, we trim a rack of three plate short ribs (NAMP 123), which are squarish in shape with heavy bones and dense, fatty, and collagen-rich flesh.

MATERIALS NEEDED:

Boning knife

Vacuum sealer and bags, plastic wrap, or zipper-lock bags for storage

1 Veal cheek seen from the side—notice that the meat is layered with connective tissue, which is tenderized during long, slow cooking, imparting richness from the collagen it contains.

2 Turn the knife so you're cutting away from your body, pulling up strips of silverskin with your other hand. Cut up and toward the silverskin rather than down and toward the meat.

3 Cut the silverskin away from the cheek until the meat is exposed on the entire upper surface.

4 Insert the knife between the silverskin and the meat on the underside, keeping the knife pointed down and toward the silverskin, and scrape it away while holding on to the edge with your other hand.

5 Trimmed veal cheeks ready to cook.

TRACY SMACIARZ:
OWNER OF HERITAGE MEATS, ROCHESTER, WASHINGTON

Tracy Smaciarz owns an artisan butchering company that specializes in locally grown and sustainable meat products. He is a charter member of the Puget Sound Meat Producers Cooperative, a nonprofit cooperative of local ranchers, farmers, butchers, restaurant owners, and others who joined in the operation of a mobile, USDA-inspected meat processing unit. Smaciarz ages his beef twelve to fourteen days or longer and can hang eighty head of cattle in quarters for dry-aging. Hunters also bring in their game, including deer, elk, moose, bear, and wild boar, to be cut and wrapped. He does the buying, aging, cutting, packaging, and marketing and also prepares cooked meats, pepperoni, jerky, and ham and bacon from his smokehouse.

I've cut up over ten thousand beef and over twenty thousand hogs in my life. I can scan a carcass in seconds and tell how good it is.

DO YOU HAVE A HERITAGE OF BUTCHERY IN YOUR FAMILY?

I am at least the fourth generation of my family to work with cattle. My great-grandfather immigrated to Minnesota from Poland around 1900 and the family moved to Pe El, Washington, in the late 1920s. He had small farms in both states and they home-slaughtered and home-processed their beef. My grandfather had about 150 acres [60.7 hectares] with about 120 cow-calf pairs. Feeding them was the highlight of my farm visits. In 1968, my dad became a meat cutter for a chain of supermarkets. Back then, they did formal training through the union. I still have his textbook, *The Meat We Eat*, which I also used to learn to cut meat. About 1975, my dad opened his own shop out of our detached two-car garage. He also owned a mobile slaughter truck, so we traveled from farm to farm, custom butchering and cutting beef, pork, and lamb. It was the epitome of the family-owned business. At a very early age, I became the cleanup kid. By five or six, I was helping make sausage. By ten or eight, I was slicing bacon. My son is ten—there's no way I'd let him do that!

WHAT MAKES YOU AN ARTISAN BUTCHER?

I still look at meat processing the same way as I did growing up, always striving to make it so attractive that it could be in a magazine article. I've cut up over ten thousand beef and over twenty thousand hogs in my life. I've cut eighteen-year-old cows, British Holstein, Hereford, Black and Red Angus, French Limousin, Tarentaise, Charolais, Indian Brahma, Piedmontese, Swiss Simmentals, beefalo, buffalo, emu, ostrich, and various breeds of pork. I can scan a carcass in seconds and tell how good it is.

HOW DID YOU BUILD YOUR BUSINESS?

Although I quit slaughtering about twenty years ago, I'm still an old-school butcher. By necessity, I'm a very observant person. You didn't ask questions of my father. He was left-handed and I'm right, so I had to learn to cut meat by observation. My dad, who was an encyclopedia of meat, was pulling flat irons in the 1970s and calling them "Western steaks." It's hard work and most people growing up in butcher shops don't continue on. In 2006, we moved in to a former bus barn, which we gutted, and put in a meat shop. We've since expanded to 6,000 square feet [557.4 square m]. The meat is slaughtered at three off-site facilities.

WHAT ARE YOUR CUSTOMERS LOOKING FOR?

In this region, people are looking for local grass-fed beef, though not necessarily certified organic. My customers may not be that interested in grade, but they want truth and transparency. They want to know who raised the animal, how it's raised, and how it's fed. I try to take the mystery out of processing. If they want to come to the plant, they're welcome. Most of the beef I sell [two to five head a week] is naturally raised and finished on alfalfa, barley, and corn from the cannery so it's more like hominy. I buy five to ten head hogs weekly from Jerry Foster, who runs a six-hundred-head dairy producing antibiotic- and hormone-free milk. The hogs eat the whey [similarly in the Parma region of Italy, the hogs eat the whey that is a by-product of making Parmigiano-Reggiano cheese].

WHAT DO YOU TEACH IN YOUR MEAT-CUTTING CLASSES?

I work with FareStart, a culinary job training and placement program for homeless and disadvantaged individuals. I took a half hog to show them the different cuts. When I do my classes, I'll break down the meat, a chef will take some of the cuts and grill them Argentinean style, I'll take the remainder and vacuum-seal and grind it, and they'll sell some cuts. In my classes at Heritage Meats, it takes me just one hour to break a side of beef down into all the cuts, using just a hand saw and a knife. I also use a miniature band saw and a portable grinder.

IS THERE MORE INTEREST NOW IN NON-MIDDLE MEATS?

More and more chefs are buying a wider range of cuts. Chuck eye steak is gaining in popularity. More people should try tri-tip, and inside and outside skirt [page 48]. Goat is becoming more and more popular, but goats are difficult to raise and they're not all created equal. If not done right, the yield ratio is horrible. I like pork sirloin for roast, chops, country-style spare ribs, stew meat, or kabobs. I'm also a big lamb person. While participating in Cochon 555, a traveling event where chefs prepare pork dishes accompanied by wine from small wineries and watch two butchers in a head-to-head cutting competition, we used the blood for sausage, and made chorizo, Italian sausage, and breakfast sausage.

VEAL BREAST

Veal breast (NAMP 313) is one of the least expensive cuts of a generally high-priced meat, and is equivalent to the beef brisket plus the plate (or belly). This tough but dense and flavorful cut consists of two main muscles, the large, lean flat and the smaller, fattier point, plus the continuation of those muscles into the plate section. Veal is rich in gelatinous collagen contained mostly in the connective tissue and the soft bones. When cooked slowly in a moist environment, the meat becomes fork-tender as the collagen breaks down and forms a rich, delicious liquid full of body. The veal breast is a good choice for braising (slow-cooking in a small amount of flavorful liquid). If stuffed, the breast of veal will serve about twelve.

> Veal short ribs may be cut from the breast, but you need a saw to cut through the bones without shattering them. Other names for veal breast include *Kalbsbrust* (German), *poitrine de veau* (French), *pecho de ternera* (Spanish), and *petto di vitello* (Italian).

1 Cut away a strip from the outer edge of the breast. Pull up the freed edge of the white membrane. While pulling with one hand and cutting away with the other, remove the membrane.

BONING VEAL BREAST
AND CUTTING A POCKET FOR STUFFING

We start with a whole bone-in veal breast (shown from the inside), remove the bones, and then cut a pocket inside for stuffing. The inner side of the veal breast is covered by a tough, thick white membrane called the *peritoneum*, which must be removed. The thin, but tasty skirt muscle, which lies over the rib bones at the larger, rear end of the breast, may either be left on or removed for another use such as grilling or pan searing.

Once the veal breast has been trimmed, you may stuff the pocket as desired, then roll up from the smaller end and tie as for a roast (see page 12). Or, stuff the breast without rolling and sew the opening closed. Either way, braise the veal breast, cooking as low and slow as possible to keep the meat moist and tender. Remove from the oven, allow the veal breast to rest for about 20 minutes and slice across the grain to serve.

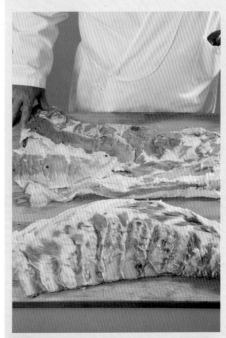

4 The boneless veal breast is nearest to the butcher. The separate rack of veal bones may be cooked like pork ribs, either whole or cut into individual ribs.

2 Cut off the underlying brisket bones, the small sternum bones that connect to the rib bones.

3 Starting at the large end, cut along the back to expose the row of rib bones. Pull away the flap of meat that covers the bones, then, running the knife along the bones, cut away the separated flap.

5 Turn the breast so its fat side is up and, using a scimitar or chef's knife, trim off most of the surface fat in thin sheets.

6 Turn the trimmed breast so the small end faces your body. Insert the tip of a butcher's knife or a slicing knife into the center and push until almost the entire blade is inside the breast.

7 Reaching in so that most of your hand is inside the pocket, move the knife in a sawing motion to enlarge the pocket without cutting through the meat.

ABOUT PORK

Pork, the meat of the domestic pig, *Sus scrofa*, is descended from European and Asian wild boars. Pig is thought to have been domesticated 11,000 years ago in central Asia. Today, pork is the most common meat worldwide, representing close to 40 percent of all meat production.

Fresh pork is highly adaptable and can be seasoned with sweet, fruity flavors like peach, ginger, and apple or savory flavors like Chinese bean sauce, garlic, and resinous herbs like thyme and savory. Modern commercial pork has been bred to be quite lean, so avoid dryness by not overcooking, especially when preparing leaner cuts like loin and tenderloin. The USDA recently lowered its recommended temperature for cooking pork to 145°F (63°C), which makes for much juicier results.

Because pork spoils easily, much of it is cured, pickled, or smoked. Due to religious prohibitions, pork is not eaten by observant Jews or Muslims.

Because it comes from a relatively small animal, pork is divided into just four primal cuts at the packing house: the shoulder, loin, belly, and leg.

Heritage breed pork including Hungarian *Mangalitsa*, Tuscan *Cinta Senese*, Spanish *pata negra*, and British Tamworth are being raised by small, sustainable farmers who often sell the meat directly or in specialty markets. The domestic pig is thought to have descended from its wild ancestor, the wild boar. Many of today's heritage breed pigs are a cross of wild boar and domestic pig. Because wild boar has the same internal structure as domestic pork, though its meat is leaner, tougher, and stronger in flavor, it is prepared the same way. Boar is considered by many, but not all, to be the same species as pork and is butchered the same way.

BONING AND TRIMMING
A WHOLE PORK LOIN PRIMAL

The long whole pork loin primal (NAMP 410) is shown at right from the inside with the short loin and its darker tenderloin section at the back. It includes the entire pig's back, starting with a portion of the chuck at the head end all the way to the sirloin at the rear. (A center-cut pork loin, NAMP 412C, has had the chuck and sirloin removed.) The whole pork loin primal includes fourteen ribs (beef, lamb, and goat have thirteen), the eye of meat (the *longissimus dorsi*—the main back muscle that stretches from chuck to sirloin), and the tenderloin, which is often removed for separate sale. In continental Europe, the pork loin primal also includes a portion of the meaty neck bones.

The whole pork loin can be cut into bone-in pork chops, and pork T-bone and sirloin chops. It is necessary to use a saw to cut through the chine (spine bones), if present. The back fat, the skin and fat that covers the loin muscles along the back, is removed for separate sale. In the UK, boneless loin is made into lean bacon (back-on) similar to Canadian-style bacon.

The rib end is the tenderest part but also fattier with more muscles surrounding the main eye of meat. The loin end is firmer and leaner and consists mostly of the large eye and the tenderloin. The sirloin end is dense and quite lean, and includes part of the complicated aitch bone, which is in turn, part of the large pelvic bone. Other names for pork loin include *carré* (French) or *Karree* (German) for the rib end, and *lomo* (Spanish and Italian) for the loin end.

MATERIALS NEEDED:
Boning knife
Scimitar or slicing knife
Sharpening steel

1 Remove the darker red tenderloin, which covers the feather bones toward the rear of the loin.

2 Grasping a boning knife in your fist as shown, cut away the meat from around the rounded tailbone, holding back the freed flap of meat with your other hand.

3 Stick the knife into the loin and cut around the rounded portion of the end of the tailbone to free the meat from the bone.

4 To remove the bones from the loin and sirloin, make a crosswise cut just past the last rib bone, cutting toward the spine.

5 Starting at the sirloin end, cut in a curve to remove the loin and sirloin meat from the bones following the shape of the backbone and feather bones, cutting toward the front end. The remaining front portion includes the rib bones.

6 Grasping the boning knife as shown for increased leverage, cut underneath the chine (spin) and feather bones on the spine side to release the loin and sirloin on the other side of the pork loin primal.

9 At this point, all the bones from the loin and sirloin end have been removed from the meat.

10 Return to the rib section and cut further between the ribs and the eye to separate the two, using the rib bones as your guide and curving your knife to hug the bones.

7 Cut the meat away from the rib section, cutting from the rear toward the front between the eye of meat and the rib bones, grasping the meat at the sirloin end with your other hand for stability.

8 Turn the pork loin as shown and finish cutting the meat away from the top end of the bones in the loin section.

11 Remove the entire bone section, including the portion of the sirloin aitch and tailbone to the rear, the spine and feather bones (sticking out sideways from the spine), and the rib bones at the front end. The pork loin primal is now boneless and ready to cook or prepare further.

Davide Fedele, *Salumaio* [producer of fresh meats] of Arcadia in the Maremma, and native to this coastal region of Tuscany, raises heritage breed Cinta Senese pigs as free-range animals in the local woodlands. It's likely that this breed, the only surviving native Italian pig, mixed with the wild boar [cinghiale] that thrive in the region. Once endangered, the Cinta Senese was recently awarded DOP status [Denominazione di Origine Protetta, protected origin] by the European Union. This black pig has a white "belt" across its shoulders, sides, and front legs, sturdy limbs, long ears that cover and protect its eyes from branches, and a long snout so it can dig in the dirt. Fedele prepares the meats himself and sells his organic salume [cured meats] directly to top restaurants. Discerning Italian chefs seek out his products, which have multilayered flavor and soft, melting fat.

DAVIDE FEDELE:
SALUME PRODUCER AT ARCADIA AZIENDA AGRICOLA BIOLOGICA, TUSCANY, ITALY

WHAT IS THE HISTORY OF THE CINTA SENESE (BELTED SIENNESE) PIG?
The Cinta Senese is an autochthonous [native] Tuscan pig that originated in the municipality of Casole d'Elsa, near Siena. The pigs appear in a fresco by Ambrogio Lorenzetti [1319–1347] called Buon Governo (good government) in Siena's Palazzo Comunale.

HOW WOULD YOU CHARACTERIZE THEIR MEAT?
The meat of the Cinta Senese is red in color and contains about 3 percent intramuscular fat providing optimal taste and flavor. It is also lower in saturated fatty acids and cholesterol than the traditional large white pig.

HOW DID YOU GET STARTED IN RAISING CINTA SENESE PIGS?
By chance, I met a businessman willing to invest money in top-quality salumi production who needed someone with appropriate knowledge and experience, like me.

WHAT KIND OF TRAINING PREPARED YOU FOR YOUR WORK AT ARCADIA?

I had my experience coming from my peasant family, then I got great help and know-how from Professor Carlo D'Ascenzi at Pisa University. I studied various preservation techniques and the use of cellar preservation at different temperatures, because my organic products are prepared without salnitro [nitrates].

HOW OLD ARE THE PIGS AT SLAUGHTER AND HOW LARGE ARE THEY?

The pigs are between twenty months and two years at slaughter and weigh between 120 and 150 kilograms [264 to 330 pounds]. Normal pigs reach 150 kilograms after eight months!

HOW MUCH USABLE MEAT IS LEFT AFTER SLAUGHTER?

That depends on the final products you want to make: The average is 60 percent.

WHERE DO YOUR CINTAS LIVE?

They live on 120 hectares [296.5 acres] of private wood: 80 hectares of chestnut and 40 of cerro [Quercus cerris, an oak native to southern Europe, also known as bitter oak and turkey oak].

WHAT DO THEY EAT?

My Cintas eat chestnuts and acorns for three months a year then grains such as barley, corn, and wild broad beans [fava beans].

DO YOU EVER USE ANTIBIOTICS?

No antibiotics are used at all. They only get vaccinations according to the law.

WHERE DO YOU GET THE PIGLETS—ARE THEY BABIES OF YOUR OWN PIGS?

The babies are only home products.

ARE THE PIGS THAT YOU USE BOTH MALES AND FEMALES?

Both female and male Cintas are butchered.

ARE THE MALES CASTRATED?

Most of the male Cintas are castrate. We leave the strongest for reproduction.

ARE THERE ANY DIFFERENCES IN THE MEAT BETWEEN MALES AND FEMALES?

Female and castrate are the same and better than pure male meat.

DO YOU EAT CINTA SENESE MEAT FRESH? DO YOU HAVE ANY FAVORITE SEASONING?

Cinta meat can be used for tagliata [cuts or chops], tartare [finely chopped raw meat], or for stuffing and roasting. The main herbs we use in Tuscany are rosemary, sage, thyme, and juniper berries.

PLEASE TELL ME ABOUT YOUR CURING PROCESS.

The procedure is rather complicated: First we do ten days of stufatura [literally "stifling," air-drying and aging] at appropriate cellar temperatures, which varies from 10°C to 19°C [50°F to 66°F] with humidity from 40 to 70 percent. During this time, special natural fermentation starts so that no chemical ingredients like nitrates are needed. A two months aging at 10°C [50°F] follows.

HOW DOES THIS AFFECT THE MEATS AND STORAGE RECOMMENDATIONS?

The salame is darker with stronger and better taste, but shelf life is shorter because organic products suffer in humidity and warm temperatures.

WHAT ARE YOUR FAVORITE SALUME TO MAKE?

Salame al miele [salami with honey] and prosciutto [air-cured uncooked ham].

WHAT IS THE MOST CHALLENGING TYPE OF CURED MEAT THAT YOU MAKE?

The most difficult one to make is my favorite salame with honey and finocchiona. [This traditional Tuscan salame is flavored with finocchio, or fennel seeds, preferably wild, and has long been featured at traditional Florentine feasts.]

WHAT IS THE MOST POPULAR SALUME YOU MAKE?

Our soppressata, which we season with spices such as cinnamon, nutmeg, and cloves. [Tuscan soppressata starts with a boiled pig's head. The meat is picked off, seasoned, mixed with the gelatinous cooking liquid, and stuffed into a large casing. It is similar to brawn and head cheese.]

CAN YOU TELL ME ABOUT THE YEARLY CYCLE OF ARCADIA?

In former times, all the preparation was done during the winter season because preservation was difficult. Now we have refrigeration and we can work all year round.

WHEN ARE YOUR PRODUCTS READY?

Our salami needs four to five months to be ready for the market; prosciutto requires a minimum of two years.

DO YOU USE AN ELECTRIC SLICER TO CUT THE MEAT OR DO IT BY HAND?

We cut it by hand. Cinta fat is quite sensitive to temperature, so using a machine to cut it melts the fat. [Spanish pata negra ham is also cut by hand for the same reason.]

DO YOU HAVE ANY SERVING RECOMMENDATIONS FOR YOUR SALUME?

Serve it with a slice of real Tuscan bread [which contains no salt] and a glass of good Tuscan red wine.

MATERIALS NEEDED:

Boning knife

Scimitar or slicing knife

Sharpening steel

Plastic wrap

Tray lined with wax or butcher paper

1 For a neater look, trim off any ragged edges, at the sirloin end.

2 Turn the loin so that the fat (outer) side faces up.

PREPARING A DOUBLE PORK LOIN ROAST

A double boneless pork loin (NAMP 413A) consists of two halves of boneless loin placed over each other with the thicker end placed over the thinner end and tied, making a larger roast ideal for groups. Here, we season the meat with sea salt and freshly ground pepper and add fresh sage leaves and rosemary sprigs to infuse the meat as it roasts.

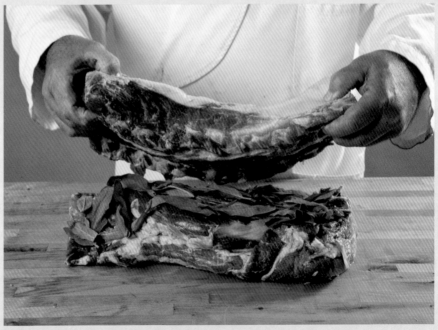

6 Season the meat and arrange a layer of sage or other fresh herb leaves (such as marjoram, parsley, rosemary, and thyme) on one side. Place one half on top of the other with the chuck at one end on top and the sirloin at the other end on the bottom so that the thickness of the roast is relatively even.

3 If available, switch to the scimitar (or use a slicing knife) and cut away most (but not all) the fat. Cut parallel to the surface of the meat and remove the fat in small sheets.

4 Turn the trimmed loin back over so the inner side faces up. Score the meat crosswise at the halfway point to divide it into two halves.

5 Check that your score is at the halfway point, then cut the loin in two.

7 If desired, place rosemary sprigs on top to infuse the meat as it roasts.

8 Tie the roast using a butcher's knot (page 12) and making the first tie in the center.

9 Tie a second string close to one end and tie a third string close to the other end.

10 Once the three basic strings have been tied, tie two more strings between each set of strings (first and middle, and middle and last) so the roast is securely and evenly tied.

11 Because this roast is large, tie a string lengthwise outside of the crosswise strings for added stability and to act as a handle when moving it from the roasting pan to serving platter.

12 Whole pork loin roast ready to cook. Because of its rounded shape, this roast is ideal for rotisserie roasting, common in France, Italy, and Greece.

MATERIALS NEEDED:

Scimitar knife or slicing knife

Sharpening steel

Butcher's string (for butterflied loin)

Plastic wrap

Tray lined with wax or butcher paper or vacuum sealer with bag

STUFFING BONELESS PORK LOIN TWO WAYS

Here, we stuff a boneless pork loin (NAMP 412B) two ways, first by cutting a cross-shaped cut in the center of the eye and second by butterflying the loin (cutting it open horizontally and keeping the two sides attached.) For stuffing, we use California tangy deep-orange-colored apricot halves soaked in warm water until plump and soft and then mixed with chopped rosemary. Turkish apricots may also be used, but these whole pitted fruits are lighter in color and less tender.

1 Cut a pocket in the loin by sticking the tip of a long thin knife through the center of the loin eye.

2 Keep the knife at the center of the loin eye while pushing it through toward the other side.

3 The knife has been inserted as far as it can go. Turn the loin eye around and stick the knife in the center of the loin eye from the other end to meet the first cut.

4 Turn the knife so the blade is vertical and stick it in crosswise to the first cut, pushing the knife in to form a cross-shaped cut.

5 Boneless loin of pork ready for stuffing.

6 Push the stuffing (here, apricots) in from one end.

9 Open up the loin and place a layer of apricots over top, using just enough to cover the surface.

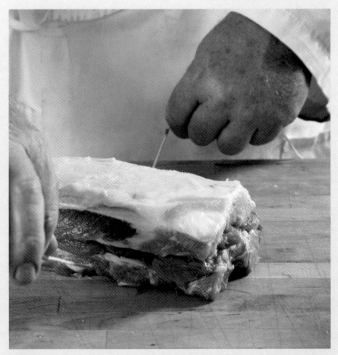

10 Tie closed with butcher's string, placing the first string at the center (page 12).

7 Turn the loin around and push the apricots in from the other end so that the loin has a center of apricots. The stuffed boneless loin is ready to cook and does not need to be tied.

8 To stuff the pork loin by butterflying, use a scimitar or slicing knife to cut horizontally about three-quarters of the way through the loin, cutting toward the outside and leaving the two halves attached.

11 Tie two more strings halfway between the middle string and the ends for a total of three. Tie a string lengthwise and outside the crosswise strings for added stability and to help lift the roast while transferring it from roasting pan to serving platter.

12 Butterflied, stuffed, and tied boneless loin of pork ready to roast.

HERB AND KATHY ECKHOUSE:
OWNERS OF LA QUERCIA, NORWALK, IOWA

Kathy Eckhouse lived in Europe for several years in her youth and worked as a ranch hand and as a researcher in agricultural economics at the University of California, Berkeley, before raising her family. Herb spent more than thirty years in agriculture, raising seed potatoes, working cattle, and developing and marketing commodity crop and vegetable seeds. Before starting La Quercia [the oak tree], their prosciuttificio [prosciutto manufactory], the Eckhouses spent five years importing prosciutto from Parma, and researching and experimenting while learning to make their own. The Eckhouses work together in all aspects of the business, selecting and buying pork, developing their own spice blends, salting, trimming, and handling hams, and leading the small, dedicated staff.

WHAT INSPIRED YOU TO START LA QUERCIA?

We spent three and a half years [1986 to 1989] in Parma, the production center of world-famous dry-cured prosciutto di Parma ham. We had never tasted really good prosciutto before that. Living in the prosperous provincial city of Parma, we became immersed in Italian culture and found an unbelievable concentration on the quality of ingredients. When we moved back to Iowa, we brought along our newfound interest in prosciutto. We were struck with the fact that this amazingly fecund and fertile land was mostly used to raise commodity crops. We were ready to try something new: making top-quality prosciutto in Iowa.

HOW DID YOU GET STARTED?

When we started making prosciutto ourselves, it was very *fai da te* [do it yourself]. We started the prosciutto in our garage in an 8-foot [2.4 m] refrigerator and hung it to age in our basement guest bedroom, using a fan to draw in fresh air. People claimed we needed Italian pork and the cool, dry breezes of the Mediterranean coming down the pine-clad slopes of the Apennines. We used very high-quality raw meat from Iowa family farms and the prosciutto was delicious.

HOW DID YOU BUILD YOUR LOCAL PROSCIUTTIFICIO?

In 2003, we bought land south of Des Moines and designed the building and setup in consultation with experts in Italy. We sourced equipment from Frigomeccanica in the province of Parma, which specializes in equipment for prosciutto production used by 35 percent of Italian producers. We opened in February 2005 and introduced our prosciutto in October 2005. Since then, we've added a slicing room and now sell some of our meats presliced. For theater, nothing beats slicing with a manual Berkel slicer or securing the prosciutto to a special stand and hand-slicing.

WHERE DO YOU SOURCE YOUR PIGS?

We have always used pigs that are nonconfinement and humanely raised without nontherapeutic antibiotics. At first, we sourced from Niman Ranch Pork in Iowa, which had no breed distinction, and the Organic Valley Coop in Wisconsin, which raised Berkshire-cross pork. It was the beginning of the education of our palates to taste differences among breeds.

We're oriented to learning, rescue, and research and now have twenty-four pigs we're raising from a farmer who is working on a special breeding project. We buy meat from market pigs [barrows—castrated males, like steers, and gilts—young females] but we can't tell which sex they were when the meat arrives. The pigs are eight to ten months old at slaughter with a live weight of 270 to 290 pounds [122 to 132 kg]. Upon arrival in our plant, the meat is identified by lot with a code that stays with each piece from the day of arrival until departure as fully cured meat.

WHAT BREEDS DO YOU USE?

We make five kinds of prosciutto: Americano is a 25 to 50 percent Berkshire male cross with a female large white pig. Green Label is certified organic from a Berkshire male on a white female. Tamworth is a mix of Tamworth with a Berkshire/Lancaster/Duroc cross. Rossa Berkshire is 100 percent Berkshire. Acorn Edition is limited-edition prosciutto from pasture-raised Berkshire and Berkshire/Chester White cross, some organic, some not. Most of our prosciutto is sold boneless, but our Acorn Edition is bone-in, hoof-on in the style of Spanish Ibérico hams. Our standard cut, known as a *sgambatto*, is minus the shank so it has a much higher proportion of usable meat [over 90 percent] than banjo-shaped Parma or San Daniele ham.

HOW ARE YOU ABLE TO GET THE HIGH-QUALITY PORK THAT YOU NEED?

Our specifications are so strict. We examine each piece repeatedly—at least fifteen times—as it goes through the curing process. Our Italian advisor said he'd never met anybody as fussy about meat as Herb, "*Non* e *nessuno pinolo comme Herberto*," in Italian. [A *pinolo* is a *Mediterranean pine nut*]. Coming from an Italian, that was quite a compliment!

WHAT ARE SOME OF THE CHALLENGES OF MAKING ITALIAN-STYLE CURED MEATS?

We nonconfinement pork, humanely raised and free of nontherapeutic antibiotics. Though Iowa is the biggest pork-producing state, we select from just 0.05 percent of its production. Another challenge is how the animals are handled at the slaughterhouse, how the killing is done, and how the carcasses are handled and chilled. All that happens before we see the meat. We have an unusual way of cutting and can accept no gashes through the skin and no penetrations through the muscle. When you dry-cure, every ding turns into a crevasse.

HOW MANY HAMS DO YOU CURE IN A WEEK?

We bring hams in every week so we have the opportunity to have the best prosciutto every week, because we can control the progression of temperature and humidity to mimic the seasons in Parma. We produce eight hundred to nine hundred legs a week [roughly forty thousand a year]. In Italy, the smaller prosciuttifici produce one hundred thousand to two hundred thousand hams a year; the larger companies produce up to ten thousand a week.

HOW DO YOU DO YOUR LABELING?

It has been hard for us to navigate the highly eroded language of meat labeling. Third-party standards can develop, evolve, improve, and worsen. Natural has no particular meaning; the terms pasture-raised and free-range don't really apply to pork. We regularly visit our farmers to ensure that we are all working towards the same goals. There's often a difference between reality and words. The same goes for restaurants. We have found that many restaurants and retailers were naming our top-of-the-line product on their menus rather than the nonbreed-specific prosciutto that they were actually buying.

HOW IS DRY-CURING DONE?

In short: put salt on, take water out, and wait patiently—it's just pork, salt, and time. This has been done for millennia—the first recipe for prosciutto that we know of comes from Cato in 200 BC. The earliest dry-cured hams were said to be made around 600 BC. We use pure American extra-coarse purified sea salt, so it doesn't just melt off. We cure all the large whole muscles on a pig including coppa [collar], spalla [shoulder], lardo [back fat], guanciale [jowls], lomo or lonza [loin], pancetta [belly], and of course prosciutto [hind leg]. Our speck [a specialty of Alpine Italy] is our prosciutto cold-smoked with applewood after curing. We use organic spices whenever possible and have no known allergens in any of our ingredients, including no gluten. Everything we make is fully dry-cured and raw, and may be eaten cooked or uncooked.

IS THERE A SEASON FOR CURING MEATS?

Traditionally, this work was done with the seasons—the cold of December for salting, the winds of January and February for riposo [resting], and the spring and summer for aging. Our building is set up on this model so that we have winter rooms, spring rooms, and summer rooms. Instead of being subject to the weather at any specific stage, which could be more or less favorable, we are able to control temperature and humidity.

WHAT ARE YOUR GOALS FOR THE FUTURE?

Working on our pork varietal program is our big focus now. The breed and the animal husbandry have a big impact on the quality and flavor of the meat, which dry-curing accentuates. We have a project in the Missouri Ozarks with a farmer who is finishing one hundred Tamworth pigs in a woodland lot where they feed on acorns. We've tried Mangalitsa [Hungarian heritage breed pig] back fat, as well as shoulder and hind leg. We have our purebred Berkshire prosciutto. We also import the beautiful back fat from Spanish *Ibérico de bellota* [heritage breed Spanish pigs that feed on acorns] and cure it in Iowa.

STEPS TO CURING PROSCIUTTO

Step 1—Salting: This is done at cold temperature—below 40°F (4.4°C)—to inhibit bacterial growth while the salt penetrates and drives moisture out of the meat. Many "uncured" meats are treated with nitrates or nitrites, either synthetic or from celery and other plant sources.

Step 2—Riposo: During *riposo* (Italian for "rest"), the meat is dried at cold temperature, allowing the salt to penetrate through the meat. This step is known as equalization.

Step 3—Stagionatura (aging): Once the meat has lost sufficient moisture, it is ready to be moved into higher temperatures where it develops its special texture, aroma, and flavor through enzymatic changes over time. Although edible, it would just be salty meat before this step.

Step 4—Trim and Package: To prevent oxidation and preserve the quality, whole muscle meats are vacuumed to remove the air and sliced meats are sealed in modified atmosphere or vacuum packaging.

CROWN ROAST OF PORK

You will need two 9- to 10-rib racks of pork ribs (NAMP 412G), preferably cut from the same animal so they match in size and shape. To make the roast, the racks are tied into a circle with ribs up and pointing out, leaving room in the center for stuffing. The rib bones should be frenched so that the ends of the bones are exposed (page 60).

The same technique can be used to make a crown roast of lamb or of veal, though their sizes will differ. The chine (backbone) must be removed from the racks at the butcher shop as it requires the use of a band saw. (Most racks of pork, veal, and lamb have had their backbones removed before sale.) Cover the exposed bone tips with aluminum foil before roasting to prevent burning. After the crown roast is cooked, the tips of the bones may be decorated with paper or foil frills, if desired.

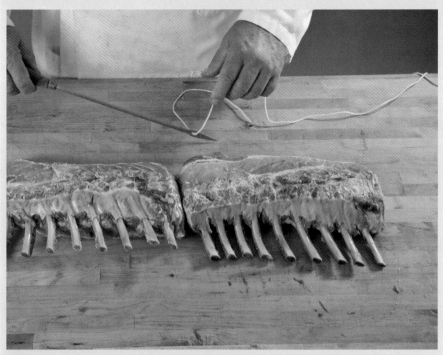

1 Start by laying the two racks next to each other with the meat-side up and with the matching ends meeting.

3 Thread the needle with more string and insert behind the first rib bone in the meaty section above the end of the bone on the first rack, exiting at the end of the first loin. Insert the needle just behind the rib bone of the second rack and exit about 2 inches (5 cm) from the separation. Remove the string, leaving it in place, then remove the needle.

2 Thread butcher's string onto the needle and run the needle from the outside through the center of the eye about 2 inches (5 cm) from the joined ends, out the end of the first section, and into the end of the second section, exiting about 2 inches (5 cm) from the separation. Remove the string and extract the needle, leaving the string in place.

4 Pull to bring together the two ends of string at the rib bones and tie with a butcher's knot (page 12).

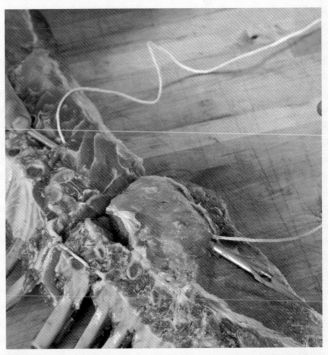

5 Insert the needle back into the meaty center section to help hold the meat in place.

6 Pull the two loose ends together and make a butcher's knot, cutting off excess string and forming a long roll of meat.

9 Remove the thread from the needle and pull the ends tight.

10 Tie with a butcher's knot (page 12).

7 Thread more string onto the needle. Curve around the two exposed ends to meet. Insert the tip of the needle into the center of the eye of the meat, about 2 inches (5 cm) from either end as you did in step 3.

8 Bring the ends together, forming the racks into a circular shape.

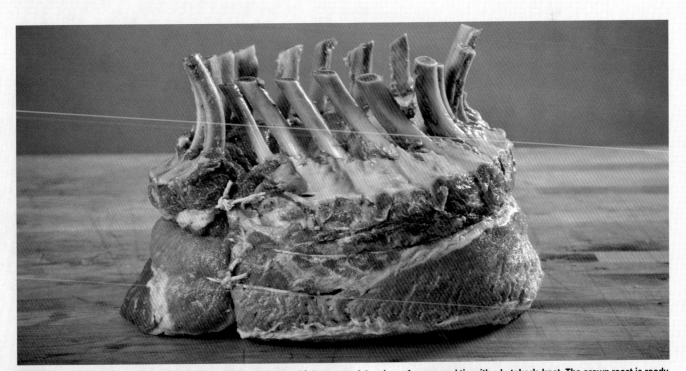

11 To strengthen the crown, run a length of string around the thickest part of the circumference and tie with a butcher's knot. The crown roast is ready to stuff and cook. You may stuff the center of the crown roast as desired, but note that a stuffing that contains meat should be precooked. Otherwise, by the time the stuffing is cooked, the meat will be quite overdone. After cooking, allow the crown roast to rest lightly covered with foil about 20 minutes, so the juices are evenly distributed and don't run out when you're carving.

A whole Jamón Ibérico de Bellota costs about $1,500... The pigs are treated like royalty. One producer plays Mozart to them, another gives them a shower every morning.

DON HARRIS:
LA TIENDA, IMPORTER OF HERITAGE BREED SPANISH PATA NEGRA HAM, WILLIAMSBURG, VIRGINIA

Don Harris grew up in a navy family, became a chaplain, and ended up in Spain, where he and his family became immersed in Spanish culture and grew to love it. When Harris retired from the navy, he and his wife settled in Williamsburg, Virginia, and decided to dress up their house with classic Moorish-style tiles. Because the tiles were very hard to find, their son Jonathan, developed a website to help others find authentic tiles. Another son, Tim, suggested adding a picture of a Jamón Serrano to their site. When the ham got an immediate and enthusiastic response, the LaTienda.com company was born and Harris became the first American importer.

WHAT GOT YOU INTERESTED IN SPANISH FOOD?
On my first navy cruise to Spain, I felt immediately at home. I loved the country and the food, which was uncomplicated, very healthy, unpretentious, and natural. After we saw the amazing response to our web notice about Jámon Serrano, we decided to start importing it.

WHY DID THE SPANISH COME TO EAT SO MUCH PORK?
One reason is that pigs are fabulous for food production so that ham is and was everywhere in Spain. Also, eating ham was a way for Christians to distinguish themselves from the Muslims who lived on the Iberian Peninsula for seven hundred years and from the Jews who first settled in Spain in Roman times.

WHAT ARE THE MAIN TYPES OF SPANISH HAM?
Jamón Serrano comprises about 90 percent of Spain's annual output and is produced from the large white pigs that flourish throughout Europe. Jamón Ibérico comes from the heritage black footed [*pata negra*] pigs that live the life of a normal pig. Jamón Iberico de Recebo is produced from pastured black footed pigs that are fed a combination of acorns and grain. For the

premier Jamón Ibérico de Bellota [*bellota* means acorns], the same pigs forage and roam in *la dehesa*, the remnants of the original forests and meadows in the south and southwest parts of Spain, for about two years. As a sign of authenticity, they are sold in Spain complete with black hooves. To comply with USDA regulations, the hooves are removed for export.

WHAT MAKES THE JAMÓN IBÉRICO DE BELLOTA SO SPECIAL?
In the fall, these favored pigs are fattened on 15 to 20 pounds [6.8 to 9.1 kg] of bellotas [acorns] per day and gain over 2 pounds [0.9 kg] daily. Finally, the pigs are sacrificed [the Spanish prefer sacrificar over the word slaughter], salted, and hung up to cure in the mountain air from two to four years, losing 20 to 40 percent of their weight in the process.

WHAT IS THE HISTORY OF THE PATA NEGRA PIG?
These pigs have been foraging and living in Spain for a very long time. There are several theories about its origins, but most agree that this unique animal is a hybrid of the wild boar, which ran wild in the Iberian Peninsula, with the domestic pig perhaps brought to Spain by the Phoenicians about 1000 BCE.

HAS THERE BEEN A REVITALIZATION OF THIS NATIVE HERITAGE BREED PIG?

Yes, Sánchez Romero Carvajal rescued the breed, which was beginning to disappear because people were plowing down the forests to build homes. With the renewed interest in the pata negra, the land is now valuable. Though other manufacturers were raising hybrid pigs, Sánchez Romero kept the breed 100 percent Ibérico.

WHEN DID THE HAMS START ARRIVING IN THE UNITED STATES?

In the past, since there were no U.S.-approved slaughterhouses in Spain, Spaniards would buy EU-approved hams from Holland and Denmark and cure them in the Spanish mountains. Eventually a U.S.-approved slaughterhouse was built and Embutidos y Jamónes Fermín delivered its first shipment of Jamónes Ibérico to the U.S. in 2007. Their premier Jamón Ibérico de Bellota followed in 2008.

JUST HOW EXPENSIVE IS JAMÓN IBÉRICO DE BELLOTA?

A whole Jamón Ibérico de Bellota costs about $1,500. What makes them so expensive is that the sows don't produce many piglets and the pigs live for one and a half to two years [American pigs average four to six months old at slaughter.] The pigs are treated like royalty. One producer plays Mozart to them, another gives them a shower every morning, yet another dispatches them painlessly with carbon monoxide.

DOES JAMÓN IBÉRICO DE BELLOTA HAVE SPECIAL HEALTH BENEFITS?

The curing process converts 60 percent of the fat of the acorn-fed pigs into beneficial good-cholesterol monounsaturated fat, much like extra-virgin olive oil.

HOW DO YOU SELL THE HAMS?

In Spain, every bar and many homes have one of these hams to be sliced off in small pieces throughout the day. To suit American lifestyles, we bring the hams in whole, slice them, and sell them in smaller portions. Americans also tend to be horrified about the harmless white mold, which develops naturally on the outside of the whole hams which we trim off.

WHAT OTHER IBÉRICO PORK PRODUCTS DO YOU SELL?

We sell lomo, loin of pork marinated in garlic, olive oil, and *pimentón* [Spanish smoked paprika]. Our *paleta* [cured, often boneless shoulders] are smaller than hams so the curing doesn't take as long. We also sell frozen Ibérico meat in four cuts.

WHAT IS THE BEST WAY TO SERVE JAMÓN IBÉRICO BELLOTA?

We serve it on a plate using a special long thin knife—like a long flexible turkey carving knife—to hand-slice the ham. Its fat melts at 70°F [21°C], so cut slices melt on the plate and in the mouth. I like it best with *membrillo* [dense quince preserve], Manchego cheese, olives, and bread. It's a staple for tapas.

MATERIALS NEEDED:

Scimitar knife

Boning knife

Tray lined with wax or parchment paper for storage

STUFFING PORK RIB CHOPS
TWO WAYS

In this technique, we cut a pocket in thick-cut pork rib chops (NAMP 1410) and stuff them with a mixture of reconstituted, drained, and chopped dried porcini mushrooms, garlic, olive oil, and bread crumbs. Pork chops are divided between the rib bones and vertebrae, and are small enough to serve as individual (or sometimes two) portions. Although they may be cut from the shoulder, loin, sirloin, or leg—in fact any portion that includes a bone—the tenderest and most desirable pork chops are cut from the rib end of the extra-long pork loin and will be moister than loin chops because they contain more fat.

1 Start with a trimmed rib end pork loin and cut chops using a scimitar knife to cut between the rib bones and the ends of the vertebrae at the base of the rib bone.

2 Two pork rib chops cut about 1 inch (2.5 cm) thick.

5 For the second method, stand a pork rib chop on its inner side with the eye facing your dominant hand. Cut a slit through the outer fat rim and into the center of the rib eye parallel to the cut side and halfway between the inside and the outside.

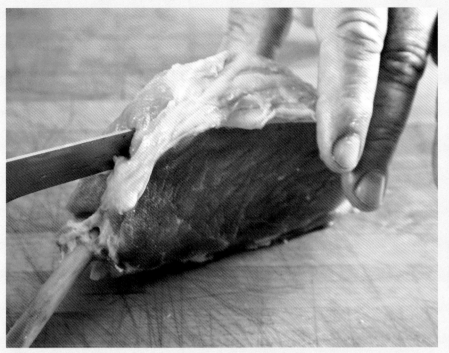

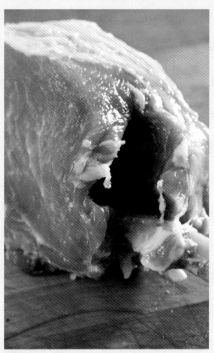

3 For the first method, stand a chop on its inner side with the bone side facing your dominant hand. Cut a slit in the rib eye parallel to and halfway between the inner and outer sides. Move the knife around inside the slit to enlarge the opening, avoiding cutting through the meat to the outside.

4 Pork chop with pocket cut close to the bone.

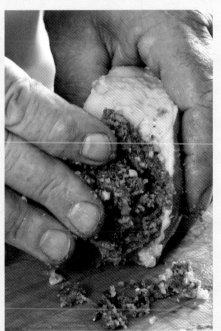

6 Move the knife around inside the slit to enlarge the opening, avoiding cutting through the meat to the outside.

7 Pack a finely chopped mixture of flavorful ingredients such as the dried porcini mushrooms seen here into the slits of the two chops, mounding the excess on the outside.

8 The stuffed pork chops are now ready for pan-searing or grilling. Finish cooking them in a preheated 400°F (200°C, or gas mark 6) oven about 10 minutes or until the internal temperature reaches 145°F (63°C) at its thickest point.

MATERIALS NEEDED:

Boning knife

Scimitar or chef's knife

Cleaver or meat pounder

Vacuum sealer and bags, plastic wrap, or zipper-lock bags for storage

PREPARING PORK TENDERLOIN PAILLARDS

The long, lean, tender, finely grained pork tenderloin (NAMP 415) is often found packed two to a package and is relatively inexpensive compared to beef tenderloin. As in other animals from chicken to beef, the pork tenderloin lies underneath the ribs running along the backbone parallel to the loin. The smaller triangular "tip," or "tail," end starts just past the rib section, while the larger "head end" is the darker reddish muscle that ends in a rounded cap-shape inside the sirloin.

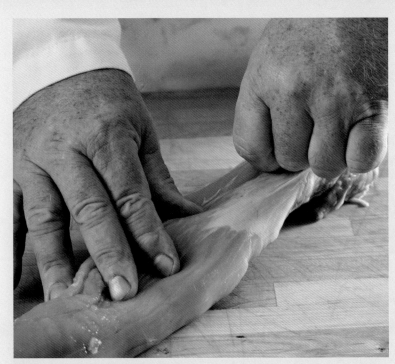

1 Pull away the loose connective tissue covering the larger end of the tenderloin.

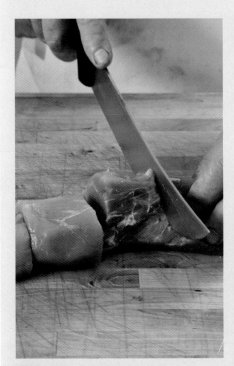

5 Cut through the tenderloin to form individual portions.

6 Tenderloin portions ready to pound.

2 Use the boning knife to cut off the underlying silverskin in thin strips; remove as little of the underlying tender meat as possible.

3 Trim a slice off the rounded tenderloin head to make a straight end.

4 Mark off three to four individual portions, making the marks farther apart as you cut toward the smaller pointed end so the portions will be of even weight.

7 Place one tenderloin portion on the work surface with its cut side facing up. Use the side of a cleaver or the flat side of a meat pounder (page 11) to pound flat. Pound three or four times, each time enlarging the diameter of the pork paillard without creating any holes.

8 Four pork paillards cut from a single tenderloin and ready to cook.

ABOUT POULTRY

Poultry is domesticated birds raised for their meat or eggs and is the second most widely eaten meat in the world, after pork. Chicken, Chinese black chicken, and turkey belong to the Galliforme family, while waterfowl such as ducks and geese are Anseriformes. The domestic chicken descended from the Southeast Asian red jungle fowl still found in the wild and probably domesticated in multiple places, as far back as 7500 BCE in Thailand. The chicken gradually traveled westward through Greece and Rome, arriving in the Mediterranean about 500 BCE and to the UK about five hundred years later. The turkey is native to North America and was first domesticated in Mexico, perhaps by the Maya. The practice of raising birds in the coop rather than allowing them to free range in the barnyard developed in Renaissance times.

In Europe and other parts of the world, it is common to buy whole chickens and other birds. Heritage breed chickens, such as France's famed Poulet de Bresse and the related California blue foot chicken, are sold whole as a guarantee of authenticity and the feet (and occasionally the head) may be served at the table.

The most common meat chicken breeds in the United States are the Cornish, which originated in the UK, and the White Rock, which originated in New England. The Araucana chicken, which has no tail, comes from Chile and lays blue eggs. The small Chinese black chicken, which has been raised there for more than two thousand years, is usually processed Buddhist-style with head and feet on. Today, pastured or free-range chickens are in demand for their firm, flavorful flesh. Whatever type or color you buy, choose poultry with a fresh, clean smell and moist but not slimy skin.

The breastbone of young birds will be cartilaginous at the rear end, as will the tips of the beaks of head-on birds (see below). Yellow or white skin color is a matter of regional or personal preference and is achieved by diet. Chickens fed with corn and marigold petals will have yellow skin. The meatiest parts of all poultry except ratites (ostrich, emu, and rhea) are the breast, or flight muscles, and the thigh and drumstick, or walking muscles.

Chicken feet make a stock rich in the collagen that gives body to reduction sauces. Some people prefer to cut off the tips of the feet to expose more of the marrow, thereby making a richer stock. Use the neck skin for helzel, an Ashkenazi Jewish sausage stuffed with flour, chicken fat, chopped innards, and fried onions. Scrape the neck skin and thread on skewers for Japanese yakatori, or roast the neck and pull off the long, tender strands of meat Caribbean style.

For chicken and turkeys that fly short distances, the breasts are light in color and the legs dark. For birds that fly long distances, such as ducks, geese, and pigeons, both breasts and legs are dark from myoglobin protein, which helps the intake of oxygen. The small, rounded "oyster" muscle on either side of the backbone just above the thighs is a delicacy with dense texture and rich flavor. Chicken wings are meaty and slightly darker in color than the breasts with full-bodied flavor. Wings of other birds such as turkey and duck contain little meat but much collagen-rich connective tissue. The legs of pheasants and turkeys contain bony, easily splintered tendons.

PREPARING HEAD-ON, FEET-ON CHICKEN

Chickens sold complete with head and feet on used to be more common, the way they are still sold in many Chinese and halal markets and at fresh-kill markets usually located in Latin American, Asian, or Caribbean communities. Buddhist-style chickens, sold in Chinese markets, are eviscerated and include the feet and heads. To qualify as true Buddhist chicken, the neck must be free of nicks or imperfections. Choosing

a perfect whole chicken is important for the Chinese New Year, because any bruises or marks can lead to possible misfortune. A "New York–dressed" chicken is sold head on and feet on but includes innards, which have been removed here.

Here, we prepare a head-on, feet-on young chicken that is halal certified.

MATERIALS NEEDED:

Sharp boning knife or chef's knife

Vacuum sealer and bags, plastic wrap, or zipper-lock bags for storage

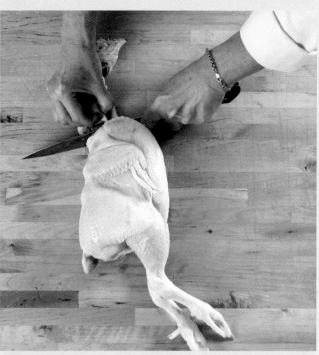

1 A head-on, feet-on young chicken killed according to the Islamic laws of halal slaughter. The animal may not be stunned and its neck must be slit after saying a prayer.

2 Using a sharp boning knife or chef's knife, cut off the neck and head.

3 To remove the head, cut between the neck vertebrae close to the body. To remove the feet, find the joint between the bones where the ankle attaches to the end of the drumstick by jiggling the bone, then cut between the bones.

MATERIALS NEEDED:

Boning or paring knife

Vacuum sealer and bags, plastic wrap, or zipper-lock bags for storage

ABOUT TRUSSING

Trussing keeps chicken and other poultry (and roasts) in a neat, tight shape. Most birds were once cooked on a rotisserie over an open hearth. If the bird was not trussed, it could flop about, burn, and throw off the balance of the turning spit, which was often turned by a mechanical device. Today, most roasting is done in an oven, so trussing is done more for cosmetic reasons than necessity. Trussing keeps the bird compact, keeps the wing tips and leg ends from burning, and makes for an attractive presentation. A trussed bird will take more time to roast, especially where the inner thigh presses against the breast, and the skin will not crisp evenly.

Trussing can be done with or without string for smaller birds, though larger birds require using string. Before trussing, with or without string, pull out any fat packets just inside the body cavity (young chickens such as the ones shown here will have little excess fat), pat the chicken dry inside and out, and season on the inside. Save the fat if desired to render for chicken fat.

TRUSSING A CHICKEN WITHOUT STRING

1 Tuck the wings behind the head and push the legs into the body, bending at the knee joint between the drumstick and thigh.

2 Pull up the loose skin between the inner thigh and the tail, stretching it taut. Cut a slit into the skin, not too close to the edge, where it will tend to rip open.

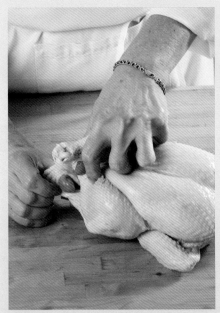

3 Pull the leg from the side opposite the slit over the other leg, then poke the end of the leg into the slit, taking care not to pull the slit open. (If this does happen, especially the first few times you try this, start over by cutting a slit into the other side of the chicken.)

4 The chicken is now in a compact shape and is ready for roasting.

TRUSSING A CHICKEN USING BUTCHER'S STRING

MATERIALS NEEDED:

Butcher's string

Scissors

Vacuum sealer and bags, plastic wrap, or zipper-lock bags for storage

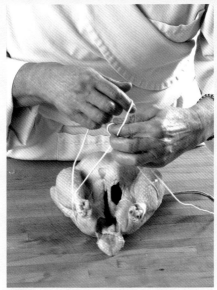

1 Cut a length of cotton butcher's string about 3 feet (0.9 m) long. Place the string under the body of the chicken and bring the ends forward outside the legs, making an X shape.

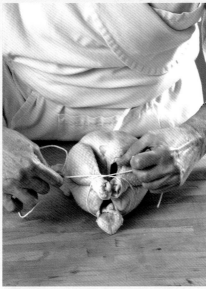

2 Drop the string below the drumsticks at their narrowest point and cross over, forming a figure eight. Pull the string tightly to close off the body cavity opening.

3 Pull the ends of the string around the outside of the body, over the legs, and over the wings.

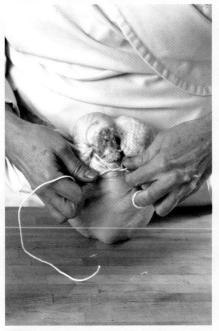

4 Bring the ends of the string under the end of the chicken neck and tie tightly with a double knot. Cut off any excess string with scissors, leaving at least 1 inch (2.5 cm) on either end to allow for slippage.

5 String-trussed chicken ready to roast.

MATERIALS NEEDED:

Boning knife

Vacuum sealer and bags,
plastic wrap, or zipper-
lock bags for storage

Glove-boning is done to remove
the entire rib cage of any bird,
basically by turning it inside out,
leaving a semiboneless bird that
looks intact and is easy to carve
when stuffed.

GLOVE-BONING A CHICKEN

Glove-boning is done to remove the entire
rib cage of any bird, basically by turning it
inside out, leaving a semiboneless bird that
looks intact and is easy to carve when
stuffed. The only place you will use your
knife is to cut through the tendons
connecting the wings to the rib cage and
the thighbone to the backbone. If you
remove the thighbones as well, which is
optional, you'll also cut the tendons
connecting the thighbone to the leg bone.

Glove-boning is suitable for smaller birds,
including game birds such as quail,
partridge, and *poussin*. The younger and
more tender the bird, the more the skin will
have a tendency to tear.

Just before cooking, stuff the bird without
overfilling, because the stuffing will expand
and the meat will shrink as it cooks.
Alternatively, season the bird inside and
out as desired, then flatten and cook under
a brick or on a grill.

To help keep the skin from
tearing on a young, tender
bird, slip a table knife (not a
sharp boning knife) between
the skin and bones. Start from
the back at the tail moving the
knife along the back of the
bird—the place where the
skin is most firmly attached to
the body—separating the skin
from the bones. Turn the bird
over and insert your fingers
between the skin and the hip
bone on either side. Loosen
the skin all the way to the front
of the bird and continue with
step 1 at right.

1 Place the chicken on the work surface breast down and facing away. Find the place where the collar (or wishbone) attaches by jiggling the wing. Cut in between the wing and collarbone to sever the tendons connecting them.

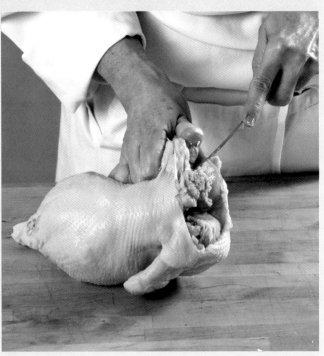

2 You should be able to pull away the wings from the neck freely. If not, cut further to sever any remaining tendons.

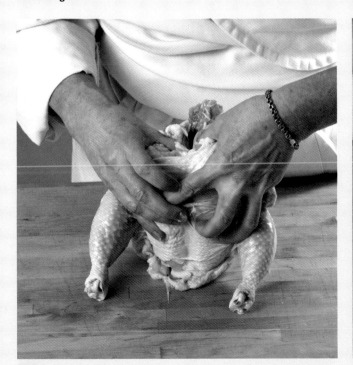

3 Pull back the flesh and skin away from the rib cage, using your hands to feel where the flesh attaches to the bones. Keep pulling back while turning the bird inside out.

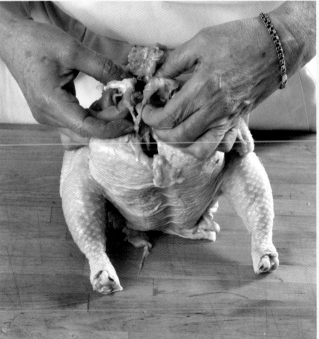

4 Remove and discard the wishbone—this bone is very fine and on a young bird as shown here, it will often break in two.

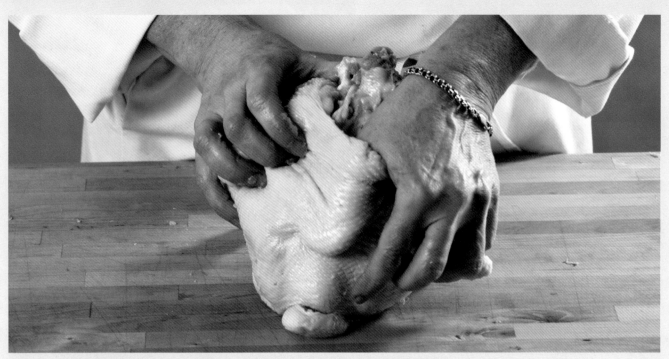

5 Use your thumb and fingers to push down between the backbone and the skin, separating the skin from the backbone with the goal of keeping the skin whole.

8 Cut the tendons connecting the rib cage to the thighbones on either side to release them.

9 Pull out the entire rib cage from the body, leaving the legs and wings intact.

6 Keep turning the bird inside out, as if you were taking off a glove.

7 Pull out the rib cage, twisting and working it gently to detach it from the body.

10 To remove the thighbone (optional), pop it out through the cavity. Scrape the meat away from the bone. Twist to release it from the leg bone and cut between the two, removing the thighbone. Repeat on the other side.

11 Glove-boned chicken shown with its two thighbones below, its wishbone at the head end, and its rib cage to the right.

Kosher meats are produced from animals of a kosher species that are processed by the body of laws of kashrut ... Large animals must have cloven hooves and must be grazing animals that chew their cud.

NAFTALI HANAU:
CO-OWNER OF GROW AND BEHOLD, KOSHER PASTURED MEATS, BROOKLYN, NEW YORK

Naftali Hanau is a *shochet* [Jewish ritual slaughterer], *m'naker* [ritual butcher], farmer, and horticulturist. He grew up in Rochester, New York, around the corner from the kosher butcher and has long been a carnivore. His paternal great-grandfather was a shochet and both sides of his family were in the hide and leather business. He founded Grow and Behold Foods with his wife, Anna Hanau, a farmer and Jewish educator who is the company's communications director.

WHAT MAKES MEAT KOSHER?
Kosher meats are produced from animals of a kosher species that are processed by the body of laws of *kashrut*. Because we don't know what many of the acceptable species of fowl are, we eat only those birds that have been eaten by tradition. They cannot be birds of prey. Guinea fowl are not eaten by Ashkenazi [Eastern European] Jews and only certain breeds, like the Emden Goose, which originated in northern Europe, are acceptable. Large animals must have cloven [split] hooves and must be grazing animals that chew their cud. They must be ritually slaughtered by a trained, licensed shochet using a special long, extremely sharp knife without nicks or imperfections, sharpened by hand with a

WHY DID YOU OPEN GROW AND BEHOLD FOODS?

Initially, I wanted to be able to slaughter poultry for myself and my family on the farm that we were living on at the time. After earning my license, my teacher required me to slaughter three birds each week. That was a lot of chickens for two people to eat and exposed my wife and me [and our friends] to the joys of eating kosher pastured poultry. I started the business after being discouraged by the lack of properly raised and delicious kosher meat on the market, especially when we realized that we weren't going to be raising vegetables.

IS THERE ANY DIFFERENCE IN THE BIRD ITSELF?

We try to get farmers to raise cockerels [young males], because their skin is a little tougher and doesn't tear as frequently. People don't want a bird with torn skin. We use breeds that grow well, are reasonable foragers, and have white feathers, because the few remaining pinfeathers will look dirty if the feathers are dark.

PLEASE TELL ME ABOUT THE ETHICAL REASONS BEHIND GROW AND BEHOLD?

I attended a Jewish day school, where I learned that kosher meat is better, cleaner, safer, and more humane because of the special rules for how Jews kill animals, which I took at face value. While doing the Adamah Fellowship in my twenties, I learned about the animal-welfare issues associated with raising large groups of animals in confinement. I became concerned about sustainability and health, and I knew I had to change my meat consumption habits. I learned to be a shochet in order to be able to shecht [slaughter] my own animals, but I realized that there was a serious need in the market for ethically produced kosher meats.

DO YOU THINK THERE SHOULD BE AN ETHICAL CONSIDERATION FOR MEATS TO BE CONSIDERED KOSHER?

It's dangerous to conflate kashrut with ethics—they are not the same thing, especially because kashrut is not about how the animal is raised or working conditions. There are now certifications including the *Magen Tzedek* [shield of justice] or *Tav HaYosher* [ethical seal] that serve as an additional mark of quality. Theoretically it's a great thing, because it provides for more accountability.

WHY DO WE HAVE THESE LAWS?

The simple answer is God told us. There are no reasons given in Torah [Hebrew Bible] for the laws of kashrut, although many commentators emphasize that they serve to teach us that we should not cause undue pain to animals. I believe that if done properly in a well-designed facility, kosher slaughter can be one of the most humane ways to end an animal's life. Killing chickens by slitting their throats is humane because it requires that slaughter done by hand, so attention to detail is higher. Our pastured birds are essentially slaughtered the same way as conventional kosher chickens—there are only so many ways you can kill a chicken. It's the production and the handling of conventional chickens that we find to be lacking.

MATERIALS NEEDED:

Boning knife

Vacuum sealer and bags, plastic wrap, or zipper-lock bags for storage

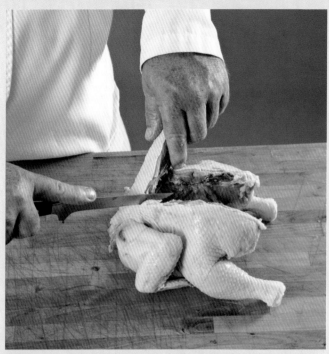

1 Starting at the tail end, cut down either side of the backbone about halfway down the length of the chicken. Pull up the freed end of the backbone, cut down either side to completely detach it, and remove.

BREAKING DOWN WHOLE CHICKEN INTO TWO HALVES

Here, we cut up a whole chicken "shell," which has had its innards removed, into two halves then remove its backbone, breastbone, and ribs. On a younger bird, such as the one shown here, the lower portion of the breastbone will be flexible white cartilage rather than bone—a good way to tell its age. Many cooks are accustomed to buying cut-up chicken, but it is useful and economical to break down a whole chicken to your own preference and use.

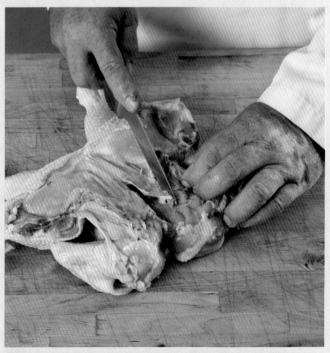

4 Remove the breastbone by cutting underneath on either side of the chicken.

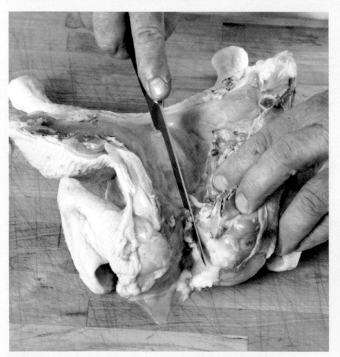

2 Turn the chicken on its breast and, starting at the head end, cut down the center of the keel, or breastbone, to about halfway down its length.

3 Grasp the chicken in your hands and crack the center of the breastbone to make it easier to remove.

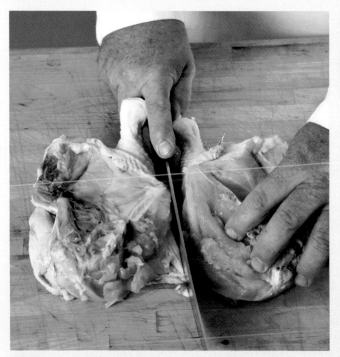

5 Open up the chicken so it lies flat and cut down the center to divide into two.

6 Pull up the edges of the rib bones and cut them away on both breasts. The chicken halves are now ready to cook.

MATERIALS NEEDED:

Boning knife

Scimitar or chef's knife

Vacuum sealer and bags, plastic wrap, or zipper-lock bags for storage

BONING A CHICKEN
LEG-THIGH QUARTER

In this technique, we start with a bone-in chicken leg-thigh quarter and remove the bones. The boneless chicken leg-thigh quarter may be cut into bite-size pieces for kabobs, grilled, or pan-seared.

1 Cut away the small bones from the outer edge of the leg-thigh quarter.

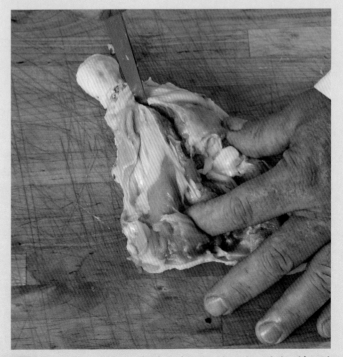

5 Turn the leg so its inner side is facing up and cut through the skin and flesh covering the leg bone as far as the ring-shaped slit.

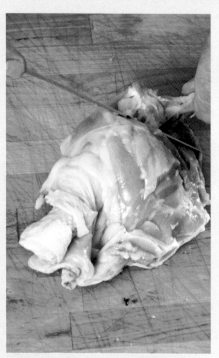

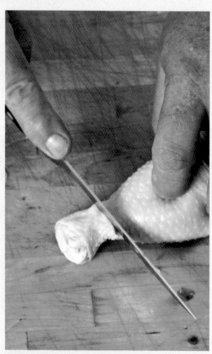

2 With the inner side facing up, find the large leg bone in the thigh portion and cut down either side.

3 Pull the flesh away from the bone and cut the bone away from the thigh meat, releasing it completely. Repeat with the second thigh.

4 To remove the drumstick bone, cut a ring-shaped slit through the skin at the bony ankle.

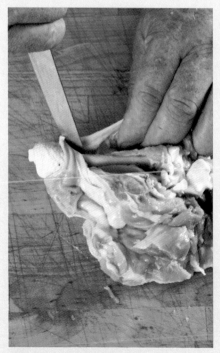

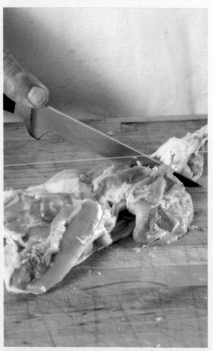

6 Cut away the flesh on either side of the bone to expose it.

7 Pull the bone out and away from the flesh and sever the tendons to cut it off. Repeat with the second drumstick.

8 Boneless leg-thigh portion of chicken shown from the inner side.

FRENCH ROLLED AND STUFFED
BONELESS CHICKEN HALF

In this French chef's technique, we remove the bones from the breast portion of a chicken breast then remove the leg and thigh meat leaving the skin attached. We butterfly (split horizontally) the breast meat so it covers the bare skin where the leg-thigh was removed. We stuff the chicken half, roll it up into a compact package, tie it, roast it, and slice it into attractive ring shapes.

Here the stuffing is roasted red peppers, sliced fresh mozzarella, and wilted spinach. Other possibilities include sautéed mushrooms with garlic and smoked gouda cheese; cooked broccoli florets and garlic bread crumbs; sliced ham and Gruyère cheese; soaked, dried apricots, sautéed shallots, and wilted spinach; or sliced prosciutto, red onion, and baby arugula.

To cook the chicken, pour oil into an oven-proof skillet then preheat. Add the chicken roll(s) skin-side down. Brown well on all sides, then finish cooking in a preheated 400°F (200°C, or gas mark 6) oven 10 minutes or until the juices run clear (not pink) when pierced at the thickest point. Remove the outer long string, leaving the underlying crosswise strings to act as a guide for slicing. Cut chicken roll into ring-shaped slices. Remove the strings and serve.

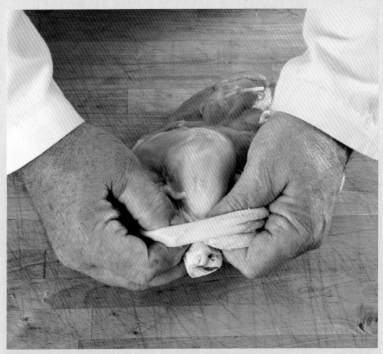

1 Start with a chicken half with the backbone, breastbone, and rib bones removed (see Breaking Down Whole Chicken into Two Halves, page 108). Pull the skin off the leg-thigh portion, keeping the skin whole.

5 Open up the butterflied chicken breast and spread it over the extra skin.

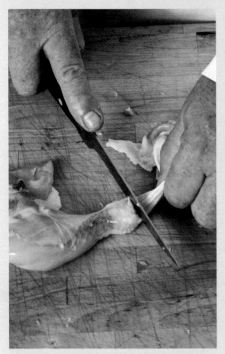

2 Remove the leg-thigh portion and cut it away, making a small hole in the skin where the leg bone pokes through.

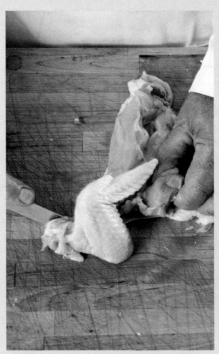

3 Cut between the second and third wing joints and remove the first and second wing joints.

4 Butterfly the chicken breast portion, cutting it in half horizontally and keeping your hand flat and parallel to the work surface.

6 Spread the filling ingredients evenly over the breast. Do not spread all the way to the edge and leave one end of the skin bare.

7 Roll up the chicken breast the short way toward the empty skin.

8 Tie butcher's string the long way around the rolled, stuffed chicken breast. Tie three sets of string crosswise around the chicken roll to make a neat, compact packet.

ABOUT TURKEY

The turkey is native to North America though it is now farmed and eaten around the world. Turkey is especially popular in Mexico, braised in complex spiced mole sauce, and in Israel, which holds the world's record for per capita consumption and where it is often layered with lamb fat and carved in thin slices for shwarma. The turkey got its misleading name from the English, who seem to have first encoun-tered this New World bird when it was imported by Turkish merchants. White meat is quite lean with mild flavor but can get dry if overcooked. Dark meat contains more fat, is stronger in flavor, but also moister, and requires longer cooking. Turkey legs contain hard, bony tendons. Smaller breasts are usually from female (or hen) turkeys and are preferred by some because they tend to be plump and juicy.

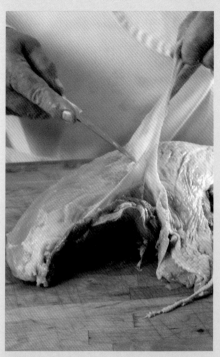

1 Place the boneless turkey breast on the work surface with skin-side up. Pull up and remove the skin.

PREPARING TURKEY CUTLETS

In this technique, we cut quick-cooking turkey cutlets and thinner scaloppine from boneless turkey breast and turkey tender-loin. If using a frozen turkey breast, it's easier to slice the thin cutlets if the breast is still partially frozen. Turkey scaloppine may be substituted for veal in any scallop-pine recipe. If the turkey has not been previously frozen, you may freeze the cutlets. Otherwise, cook within two days; once cut, the turkey will not keep well.

MATERIALS NEEDED:

Boning knife

Scimitar or chef's knife

Wax paper

Vacuum sealer and bags, plastic wrap, or zipper-lock bags for storage

Tray

5 Butterflied turkey tenderloin cutlets ready to cook.

2 Pull up and remove the loose membrane covering the rear portion of the breast.

3 Pull the tenderloin out from each breast.

4 Cut halfway through the tenderloin to butterfly the meat, then open it up to lie flat.

6 Place the main breast muscle on the work surface with the outer side facing up and the fatter end toward your body. Cut thin angled slices against the grain of the meat.

7 Continue slicing until the entire breast has been cut into thin cutlets. These are ready to cook or pound to thin them further for scaloppine (see Preparing Pork Tenderloin Paillards, page 96). Layer the cutlets between sheets of wax paper and store in vacuum or zipper-lock bags placed on a tray to catch any drips. Use within two days or freeze up to one month.

ORGAN MEATS

Organ meats, variety meats, or offal (from off fall, because it would fall off the table during butchering) refer to the internal organs of an animal that has been butchered for food. Depending on the culture and the particular organ, organ meats may be regarded as delicacies or as inedible waste. Those that come from young animals, especially veal and lamb, are held in high regard as their flavor is milder and their texture firmer. Soaking organ meats in water mixed with salt and/or vinegar will mellow their flavor. Kidneys, tongue, liver, sweetbreads, and other organ meats are quite high in cholesterol and saturated fats.

Pork organ meats are a staple of French charcuterie, German sausage-making, and Italian salumeria. Duck and goose foie gras and veal and lamb liver and sweetbreads fetch the highest prices. Scottish haggis, Italian and Mexican tripe stew, Pennsylvania Dutch stuffed braised maw (stomach), Jewish chopped liver and pickled tongue, African American chitterlings and pigs feet, French andouille sausage and sautéed kidneys, Indian braised lamb trotters, Greek kokoretsi, lamb or goat intestines wrapping seasoned mixed offal, Lebanese kasbeh nayeh, raw lamb's liver, and Caribbean souse, or head cheese, are all traditional organ meat preparations.

In France, small birds including poussin (very young chicken) and pigeon cooked in a pig's bladder (*en vessie*) is a great delicacy. In Peru, Chile, and Bolivia, *anticuchos de corazon*, grilled beef heart skewers, are popular street fare. Today, many chefs are using their knowledge and creativity to incorporate long-disdained organ meats into their cuisine. Ethnic grocery stores are the best place to find a variety of organ meats, which are often sold frozen because they are so perishable.

ABOUT LIVER

The smooth, rounded semirectangular veal liver (NAMP 3724) is one of the widely appreciated of organ meats and the largest organ in the body after the skin. Its many fans appreciate its mild flavor, fine-grained, tender texture, and high iron content. Calves, or calf's, livers are larger and darker in color, veal livers are smaller and lighter. Pork liver, often included in French pâtés, lamb liver, and chicken livers are also available. Veal liver is often cut into thin, quick-cooking slices and then sautéed as in *fegato alla Veneziana* (Venetian-style liver). Like all organ meats, liver is quite perishable, especially once trimmed and sliced. Other names include *foie de veau* (French), *higado de ternero* (Spanish), *fegato di vitello* (Italian), and *Kalbsleber* (German).

Livers that are marked kosher are highly sought after by chefs because they contain no cuts or blemishes. Visible on the left page is a USDA inspection stamp. On a kosher liver, another special stamp would be seen on the upper side as well.

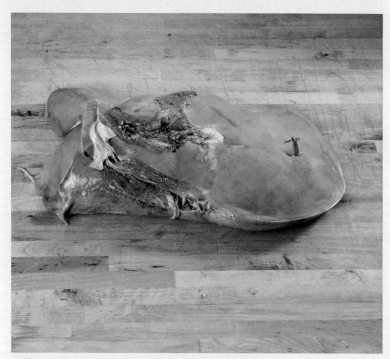

1 The whole veal liver is shown here from its underside, which contains a network of smaller and larger rubbery blood vessels that will mostly be removed.

PREPARING VEAL LIVER

The lightly mottled pale pink to light purple liver shown here comes from a milk (or formula)-fed calf. It is covered by a thin transparent membrane, which we will remove. The veal liver contains one small and one very large lobe. A large duck or goose foie gras (page 129) may be cleaned the same way.

MATERIALS NEEDED:

Boning knife

Scimitar or slicing knife

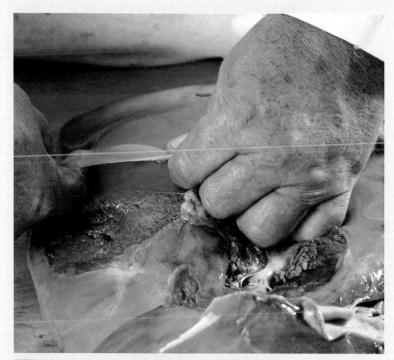

2 Make an opening in the thin membrane covering the liver and reach under with your fingers to separate it from the liver itself. A very fresh liver will be more difficult to peel.

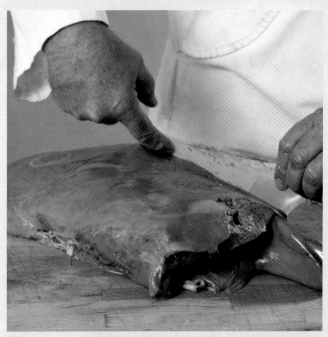

3 Detach the membrane by sticking your fingers down between it and the liver, pulling the membrane up and away.

4 Most of the membrane has been removed from the top of the liver. It is not necessary to remove every last bit.

7 Turn the cleaned liver so its underside is up. To slice, switch to the scimitar (or a slicing knife) to be able to cut larger slices and cut off the smaller lobe.

8 Keeping your other hand flat on the surface of the smaller lobe, cut the lobe horizontally into two or three thinner slices.

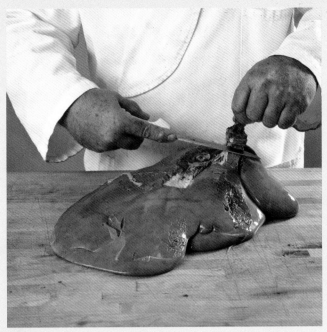

5 Turn the liver over to its underside and cut away any larger blood vessels, visible as open tubes, from the underside of the liver. The goal is to remove as many of the tough, rubbery blood vessels as possible while leaving the liver itself as whole as possible. Some smaller veins will inevitably be left.

6 The cleaned veal liver shown from its underside ready to slice.

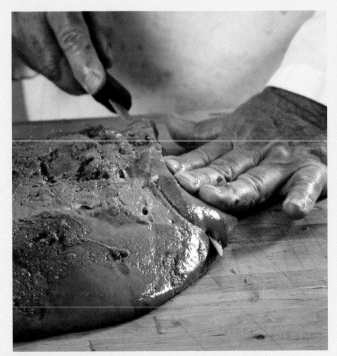

9 Cut the larger lobe across the liver at a slight angle into large thin slices. A long-bladed knife is best. If you don't have a scimitar or slicing knife, use a chef's knife.

10 Continue cutting slices of liver. Wrap and store for 1 or, at most, 2 days or vacuum-seal slices either individually or two or three at a time and freeze.

ZANE CAPLANSKY:
OWNER OF CAPLANSKY'S
DELICATESSEN, TORONTO, CANADA

Zane Caplansky, a Toronto caterer and lifelong Jewish deli lover, first set up shop in a tiny kitchen at the back of a bar, where he served Montreal-style hand-carved smoked meat sandwiches—an instant hit. Not long after, he opened Caplansky's, where he cures, spices, smokes, steams,

WHAT IS DELI? WHAT IS ITS HISTORY?

Deli was born in North America out of a need to feed the large population of Jewish working-class immigrants from Eastern Europe. It's food that was eaten at home in the Old Country including corned beef, pastrami, smoked meat sandwiches, chopped liver, and chicken soup. Delis, as we know them in North America, didn't exist in Europe; they're not throwbacks. Partly because of a history of persecution, partly because many Jews kept kosher, the community tended to be quite insular. The deli became part restaurant/part community center: a gathering place, a place to celebrate, a place to mourn, and a place to kvetch [complain].

DID YOU LEARN TO COOK THIS FOOD FROM YOUR FAMILY?

Both my grandmother and mother were wonderful cooks. I'm proud to use my mother's name for the chopped liver that I serve, which is made from her recipe, using beef liver, onions simmered in our home-rendered schmaltz [chicken fat, essential to the flavor of so many deli foods], boiled eggs, salt, and pepper. My grandmother inspired our "Battle of the Bubbies" [grandmothers] with a Rosh Hashanah [the Jewish New Year in autumn] matzo ball competition in the cannonball and fluffy categories, a Chanukah [the winter feast of lights] competition with a potato latke-palooza, for round two, and a gefilte fish competition for Passover [in spring] for round three.

WHAT MEATS DO YOU PREPARE AT CAPLANSKY'S?

We do smoked beef brisket using the double brisket, which includes both the lean plate and the fatty deckle. We brine and smoke turkeys in-house, make our chopped liver, and make pickled tongue, which has turned out to be a big seller. We also make barbecue braised beef in a thick, tomato, sugar, and vinegar-based braising liquid.

HOW DO YOU MAKE YOUR SMOKED MEAT?

We cure it by hand, applying our own custom-made salt-and-sugar dry cure mixed with our own spice blend. The salt cure draws the moisture out of the meat and is drawn back into the meat during the curing process, so the spice flavors are inculcated into the meat itself. After ten days, we rinse off the cure, reapply the spices, and hot-smoke the meat over Canadian sugar maple for eleven hours. Then, we cool the meat, trim it a bit, and steam it for three hours. The steaming helps break down the connective tissues in the brisket, a dense, tough, but very flavorful cut.

HOW ABOUT THAT PICKLED TONGUE?

We brine beef tongue for a week in a wet brine to cure it, then boil it with onions, garlic, cinnamon, cloves, and pepper until it's beautifully tender. Toronto has a very strong nose-to-tail movement and it's a meat lover's city, so our customers are especially eager to try less common cuts like tongue and chopped liver.

AND YOUR SMOKED TURKEY?

To make our smoked turkey, we wet-brine it for two days in a mixture of salt, sugar, citrus, pepper, garlic, and onion, then hot-smoke it over sugar maple, and hand-carve it for sandwiches.

WHAT IS YOUR PERSONAL FAVORITE?

On my menu, the sandwich called "What Zane Eats" is grilled verst [kosher style salami, which is cooked, not raw like Italian salami] on rye with chopped liver, red onion, and honey mustard.

WHAT DO YOU WISH MORE OF YOUR CUSTOMERS WOULD ORDER?

Chopped liver. It's profitable, tastes so good, and it's got my mother's name on it!

WHAT IS YOUR DREAM FOR CAPLANSKY'S?

I am launching a food truck, which is the only form of expansion I see myself doing. I want to have one great restaurant and catering business. If I could be as happy as I am, doing what I'm doing now for the rest of my life, there's nothing better. I love to be in the restaurant. I want to pinch the cheeks of my customers' children and see them through the different stages of their lives. My job is the furthest thing from thankless: My customers thank me just by coming and eating Caplansky's food.

ABOUT SWEETBREADS

The fanciest and most expensive of organ meats, soft pale pink veal sweetbreads (NAMP 3722), sold in pairs, is the culinary term for the two-part multilobed thymus gland found in young animals. (Lamb sweetbreads are also delicious but hard to find in the United States. Pork pancreas is similar in texture.) There are two parts of a set of sweetbreads. The more desirable side, often known as the heart sweetbread, is plump and rounded with firm, creamy texture; the other side, often known as the throat sweetbread, is longer, narrower, more irregular in shape, and contains more connective tissue. Lamb sweetbreads are much smaller but also prized, especially in the UK and continental Europe. Other names for sweetbreads include *ris de veau* (French), *mollejas* (Spanish), *animelle* (Italian, may also refer to testicles), and *Bries* (German).

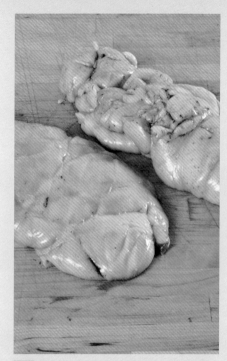

1 The throat sweetbread is to the right, the heart sweetbread to the left. Sweetbreads are covered by a thin membrane that is removed after precooking.

PREPARING VEAL SWEETBREADS

Here, we prepare veal sweetbreads by soaking them in water to mellow their flavor, poaching them in acidulated water to lighten their color, and then weighting them for firmness. The sweetbreads are covered in a transparent membrane that is removed after poaching. Like all organ meats, sweetbreads are highly perishable and are often sold frozen. Defrost in the refrigerator on a tray to catch drips. Once weighted, the sweetbreads are further cooked, usually by sautéing, grilling, or browning and then simmering in a rich sauce.

MATERIALS NEEDED:

Tray covered with wax paper, parchment paper, or plastic wrap

Plastic wrap

Half lemon

Salt

Bowl

Water

Heavy weight such as a can or a Pyrex bowl

4 Cover with a heavy weight, such as a Pyrex bowl or a can of tomatoes, to compress and firm the sweetbreads. Refrigerate overnight.

2 Place the sweetbreads in a bowl of cold water and leave to soak, refrigerated, for 2 to 3 hours, then drain and discard the water. Place the sweetbreads in a pot, cover with cold water, add a half lemon (squeeze the juice into the water), and salt, and bring to the boil. Reduce the heat, simmer 5 minutes or until the sweetbreads are firm, then drain and cool.

3 Place the sweetbreads on a tray covered with wax paper, parchment paper, or plastic wrap and cover with plastic wrap.

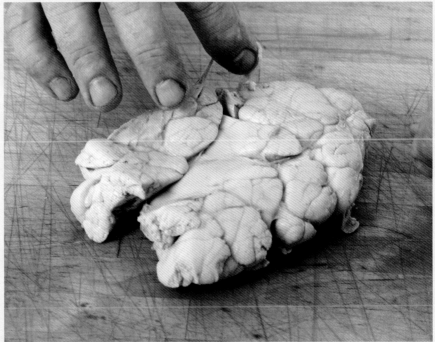

5 Pull off and discard the outer skin and any hard connective tissue.

6 Before cooking, usually by sautéing, braising in rich sauce, or grilling, slice the sweetbreads or pull them apart to make bite-size nuggets.

ABOUT KIDNEYS

There are two multilobed kidneys in each calf or cow, concave on the outer side and convex on the inner side, which has a central core of hard white fat. Very fresh kidneys will be reddish tan in color, plump and glossy and if from a young animal, they will be mild in flavor with firm texture and a faint odor of ammonia, if any. Beef kidneys are larger, darker, and more pronounced in flavor and aroma. Like all organs, kidneys are highly perishable. Mild lamb kidneys are smaller, single lobed, and dark reddish brown in color like a larger kidney bean. Stronger-tasting pork kidneys are also single lobed and dark red in color. Other names include *rognon* (French), *riñones* (Spanish), *rognone* (Italian), and *Niere* (German).

PREPARING VEAL KIDNEYS

MATERIALS NEEDED:

Boning knife

Vacuum sealer and bags, plastic wrap, or zipper-lock bags for storage

Here, we prepare fresh veal kidneys (NAMP 3728), which lie on the inside of the loin section. A traditional British veal kidney chop was made by cutting straight through the veal carcass to make chops containing a section of kidney, located at the top of the loin. Today, the veal kidney chop is formed by wrapping the tail of a loin chop around a separate piece of kidney.

1 Turn the kidneys so the inner side faces up. Use the tip of a boning knife to cut out the hard white fat core of each kidney.

2 Remove the fat, leaving a hollowed-out core in each kidney.

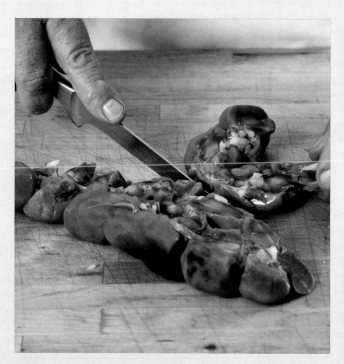

3 Cut apart the lobes at their natural seams and then slice if desired.

4 Small lobes of prepared veal kidney ready to cook. Store covered and refrigerated and cook within 1 to 2 days or vacuum seal and freeze.

ABOUT TONGUE

The long, curved veal tongue has a raised hump-like portion toward the rear, which is more prominent on the larger, darker beef tongue. The front of the tongue consists of dense muscle rich in fat and collagen covered with a thick outer layer of skin that must be removed after cooking. All tongues, including veal, beef, lamb, goat, and pork, are quite firm and noticeably grainy in texture. They require long, slow cooking though tongues of younger animals, such as veal and lamb, will be smaller and more tender. The strong smell that develops while cooking tongue will dissipate after the skin is removed. For beef tongue, once poached in liquid, the outer skin may be gripped and peeled off. For milder-tasting veal tongue, the skin is often firmly attached and must be trimmed off with a knife. The USDA does not permit the harvesting of tongue (and other organs) from cattle over thirty months old.

Like other organ meats, tongue is quite perishable and for that reason is often sold frozen. In Mexico, tongue fills tacos. In Jewish cookery, tongue is pickled and sliced for deli sandwiches (page 121). In Sichuan, China, pork tongue is cooked, sliced, and seasoned with dark sesame oil. Look for tongue in markets serving Latin American, Indian, Chinese, and Jewish populations. Other names include *Zunge* (German), *langue* (French), *lengua de res* (beef tongue in Latin America), and *lingua* (Italian).

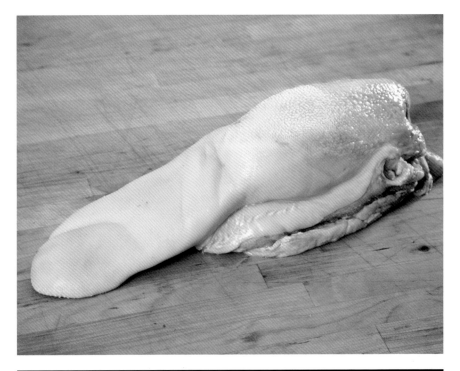

PREPARING VEAL TONGUE

A "Swiss-cut" tongue (NAMP 3710) has had the group of small bones and fatty tissue that lie underneath and toward the rear removed. If still present, this portion must be trimmed off.

Wash the tongue well, then soak in cold water 1 to 2 hours and drain. Bring a pot of salted water to boil, add the tongue, and bring back to boil. Reduce heat to low and simmer 15 minutes. Remove from the water and rinse the pot. Fill the pot again with water and the aromatic vegetables. Bring to boil, simmer 15 minutes, then add the tongue and bring back to boil. Reduce heat and simmer 45 minutes to 1 hour for tongue and up to 2 hours for beef tongue or until tender when pierced. Drain and cool.

MATERIALS NEEDED:

Boning knife

Medium to large pot

Salt

Aromatic vegetables (2 carrots, 2 ribs celery, 1 onion, 3 cloves garlic cut into large chunks, several sprigs herbs such as parsley, thyme, or dill), cut into large chunks

Vacuum sealer and bags, plastic wrap, or zipper-lock bags for storage

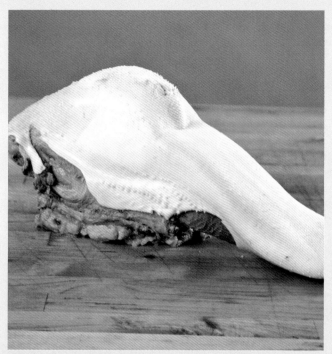

1 The skin of the tongue will turn white and opaque as shown.

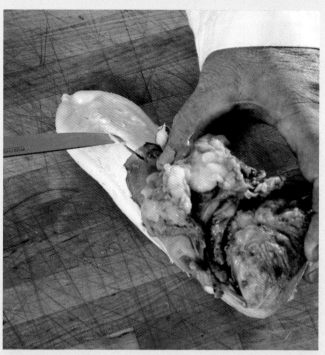

2 Turn the tongue so its underside is up, and cut a slit down the center through the skin.

3 Use the tip of the knife to separate the edge of the skin. Using your fingers, peel off as much skin as will come loose.

4 Use a boning knife to slice off the remaining skin, removing as little as possible of the underlying meat. Turn the tongue over and cut away the fatty tissue and small bones (if present) at the base of the tongue.

5 Trimmed tongue ready to further cook, often by pickling, stewing, or braising.

ANISSA HELOU:
COOKBOOK AUTHOR, TEACHER, AND EXPERT IN OFFAL, LONDON AND BEIRUT

Anissa Helou was born in Beirut, the daughter of a Syrian father and a Lebanese mother, and educated there at a French convent school. Though she is a professionally trained, world-class art historian, Helou always had a strong interest in the food of the Levant and is the author of *Lebanese Cuisine, Street Café Morocco, Mediterranean Street Food, Modern Mezze,* and *Savory Baking from the Mediterranean.* In 2005, *The Fifth Quarter*, a pioneering book on the uses and delights of offal, was published. Because of strong interest in offal among chefs and consumers, a new edition of *The Fifth Quarter* was released in 2011.

WHAT GOT YOU INTERESTED IN WRITING A WHOLE BOOK ABOUT OFFAL?
My agent advised me not to include a chapter on offal in my *Lebanese Cuisine* book and even though I listened to her, it bothered me that I had not included recipes for such an important part of our culinary tradition. Also, I love offal.

IS OFFAL IMPORTANT IN THE CUISINE OF THE EASTERN MEDITERRANEAN (SYRIA AND LEBANON)?
Yes, it's very much part of the no-waste philosophy we have, not to mention that it is considered a delicacy given how little offal there is in a whole animal, and how delicate some of the textures are.

WHAT OFFAL DID YOU GROW UP EATING?
All of it, from head to feet to tripe to raw liver for breakfast.

DO YOU HAVE PREPARATION RECOMMENDATIONS FOR ORGAN MEATS IN GENERAL?
You have to be superclean with them and in some cases, like with brains and testicles, very gentle in their handling. Of course you mustn't overcook them as you ruin the delicate textures that make them so appealing. And in other cases, like with tripe, head, and feet, you need to cook them for a long time to break down the toughness of the meat.

WHAT ARE THE SPECIAL CHALLENGES IN WORKING WITH ORGAN MEATS?
Buying them well prepared. Tripe needs to be cleaned very well by the butcher before you can take it home and wash it again in soap and water. I remember once buying lamb's tripe in a supermarket in west London. I was thrilled to have found some but wondered why it was black. When I took it back home to my mother to use it for pieds et paquets [Provencal French specialty of lamb's trotters and tripe packets] she refused to deal with it saying it would take hours to get rid of the dirt, which was why it was so black. We ended up having to throw it away. Or in the case of more delicate offal like brains or sweetbreads, cleaning them of the blood without spoiling them.

WHAT ARE A FEW OF YOUR FAVORITE ORGAN MEATS?
I love testicles and brains because of the soft, delicate texture, but I also love tripe, especially when it is stuffed the way the Lebanese prepare it. I also love trotters [feet] for their gelatinous texture.

DO YOU HAVE SOME SPECIAL SEASONINGS THAT YOU LIKE TO USE WITH OFFAL?
Not really. Each calls for a different preparation and seasoning, although I quite like to squeeze a little lemon juice on sautéed brains, sweetbreads, or testicles. Also adding lemon juice and garlic to the broth in which stuffed tripe is cooked.

CAN YOU RECOMMEND A FEW "STARTER" ORGANS FOR PEOPLE TO TRY WHO ARE NEW TO THEM?
Well, foie gras and caviar would be the obvious choice, as would normal liver. Ox cheeks is another good starter organ meat.

DO PEOPLE IN THE UK GENERALLY EAT A LOT OF OFFAL?
Kidneys and liver are very much part of traditional English cooking. You can buy liver, kidneys, even tripe in most English supermarkets.

HOW DO YOU COOK A SHEEP'S HEAD AND HOW DO YOU EAT IT?
You can boil or bake it and you eat it with your hand, picking the meat off with your fingers then opening the head [which would have been cracked by the butcher] to extract the brain, which is the prized part.

Grade A duck foie gras shown from the outer side

ABOUT FOIE GRAS

Foie gras, meaning "fat liver," is the luxuriously rich, fattened liver of duck or goose produced by fattening the birds using a feeding process known as *gavage* in which the birds are force-fed through a tube, imitating their natural tendency to force-feed before a long flight. Foie gras is high-priced special occasion fare that was enjoyed by the ancient Egyptians as well as the Romans. Goose foie gras became a specialty of Alsace, France, where Jewish residents fattened their geese to obtain their fat for cooking. Gascony is famed for its duck foie gras often from large hybrid female Pekin and male Muscovy ducks known as Moulard ducks.

Close to 80 percent of world foie gras production is in France. Hungary is also a major producer, mostly for export to France. Duck (but not goose) foie gras is produced in New York and California, (though as of 2012, it is illegal in California), and in Quebec, Canada. Larger and more expensive goose foie gras is silky in texture, with rich, subtle flavor, and is preferred for terrines; smaller, less costly duck foie gras is firmer with earthier flavor and is preferred for pan searing.

For duck foie gras, grade A is smooth and creamy, firm and without bruises. Grade B livers are softer and may show a few dark spots from bruising. Grade C livers are smaller, softer, and contain less fat. American foie gras is oval shaped; French foie gras is more rectangular in shape.

ABOUT GAME

Game is wild animals hunted for food or sport, whether furred or feathered, and includes small game such as hare and squirrel, large game such as venison and boar, and birds such as partridge and pigeon. The type and range of animals hunted for food varies throughout the world depending on climate, the local animal population, custom, and other factors such as religious beliefs. Wild game animals may be farm raised or hunted on game preserves. In many places, including the United States, wild game may be hunted for personal use but may not be sold for commercial use. However, wild game birds including grouse, wood pigeon, and pheasant from the Scottish Highlands are exported to the United States in season and are legal for sale.

Farm-raised game lives in confined outdoor areas and often fed grain, so the flavor is mild and less distinctive, though also more tender. Ranch-raised game animals are free to roam and forage, so the flavor of their meat is closer to that of wild game but also tougher. Many game animals have close domestic counterparts including wild turkey, wild goose, wild duck, bison, boar (the same species as domestic pig), wild rabbit, and hare.

DUCK AND GOOSE

Rich dark duck meat has long been a favorite in China where the White Pekin duck was domesticated at least two thousand years ago. Duck arrived in North America with the Spanish explorers. Lean Muscovy ducks were domesticated by the indigenous peoples of Brazil more than two thousand years ago and are Europe's most popular breed. A Muscovy gets its name from the musky flavor of the meat in mating season. Between 2000 and 2009, world duck production increased by almost 1 million metric tonnes (1.1 million tons). Most birds at the market are young ducklings six to eight weeks old. Larger roaster ducklings are older but still fewer than sixteen weeks old.

Geese, which are similar in structure though much larger than ducks, were bred in ancient Egypt, China, and India. The white Embden goose, of German origin, is farm raised in the United States under cover for the first six weeks and then on pasture for fourteen to twenty weeks. Ninety-four percent of world goose production is in China. Poland is the world's leading exporter of goose meat, mostly to Germany.

Although duck and goose are quite fatty, unlike beef, their fat, which helps keep them buoyant while swimming, is mostly on the outside rather than intramuscular. Their breast meat is finely grained and quite lean under the fatty skin; their leg meat is tougher and contains bony tendons. The wings have little meat though lots of collagen.

Both duck and goose parts, especially the legs, wings, and innards, are preserved by salting, smoking, and curing. They may be force-fed to produce ultrarich, smooth, and creamy foie gras (page 129).

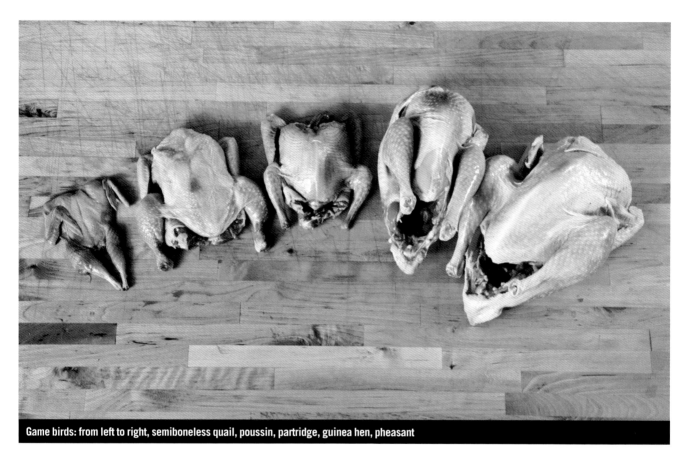

Game birds: from left to right, semiboneless quail, poussin, partridge, guinea hen, pheasant

GAME BIRDS

Quail are the smallest European game bird and belong to the Phasianidae family along with partridge, pheasant, and chicken. The common quail is native to Europe. The Japanese Coturnix quail is a popular farm-raised game bird in the United States. Quail are tender and succulent with deep pink to light brown meat and fragile bones. Because they are quite small, quail are traditionally eaten with the hands. Allow two quail for a main course, one for an appetizer.

Poussin are very small, young chickens with delicate, tender, moist meat and get their name from the French word for unfledged. Because they are so young, these birds haven't developed their flight feathers. Poussin are often glove-boned (page 102). Allow one poussin per person.

Partridges are a group of medium-size, stout, nonmigratory birds native to Europe and closely related to pheasant and quail. Their medium-dark flesh has full-bodied, earthy flavor. Similar American birds such as ruffed grouse and bobwhite quail may also be known as partridge. Spanish and southern French red-legged partridge, also found in Scotland, and the British gray-legged partridge are the most common European partridges. Allow one partridge per person.

Guinea hens, or guinea fowl, are a group of birds similar in size to chicken and native to Africa and Madagascar with reddish-colored, mildly gamy meat. They were first domesticated by the ancient Greeks and Romans and are especially popular in Italy, where they are known as *faraona*, recalling their importance in ancient Egypt. Their

French and Spanish names, *pintade* and *pintada*, mean painted or speckled. Unlike many other game birds, Guinea hens do not have bony leg tendons. Allow one guinea hen for two.

The pheasant is closely related to the chicken and has distinctive long, colorful tail feathers in more than fifty breeds and is highly regarded for its mild but flavorful flesh. The smaller hen pheasant is preferred over the cock for its fine-grained, moist flesh and delicate flavor. Free-range pheasants that live outside under nets and eat a diet similar to that eaten in the wild will be richest in flavor. Allow one larger pheasant for two to three, one pheasant per person for smaller birds.

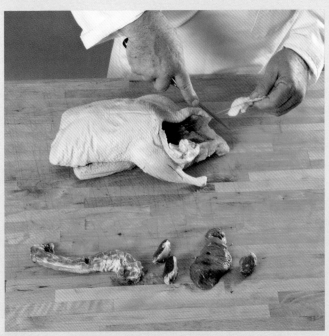

1 Pull out the fat packets from just inside either side of the body cavity. Save if desired for rendering. Pull out the neck and organs (the neck is to the left, next is the gizzard, or stomach, the liver, and heart).

BREAKING DOWN WHOLE DUCK

Here, we break down a whole duckling: its breasts to cook like a steak, its legs to preserve as confit (page 136), its fat to render, its liver and gizzards for sauces or preserving, and its carcass for stock. The same technique may be used for goose. Although institutional buyers can find duck parts in almost any form, home cooks will usually find frozen whole ducks except for the holiday season when fresh birds are in demand.

If necessary, defrost the duck or goose 2 to 3 days refrigerated in a pan to catch drips. Unwrap, drain well, and pat dry inside and out. Place the duck (or goose) on the work surface breast-side up.

4 Using a boning knife and cutting from the head end, cut through the wishbone. Then, using your hands, crack the breastbone in half, pressing up and moving your thumbs outward from underneath the breast bone.

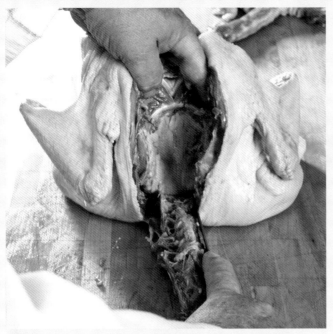

2 Cut down either side of the backbone starting at the head end with a boning knife, cutting through the small bones that attach to the backbone. Alternatively, use heavy kitchen shears or the corner of a cleaver.

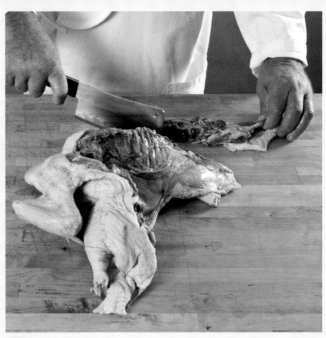

3 Open up the duck and cut away the backbone completely.

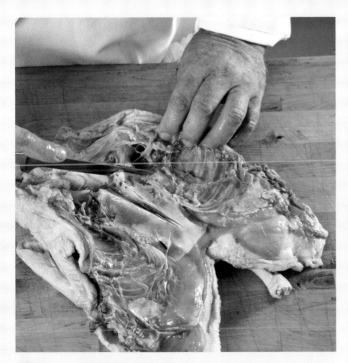

5 If necessary, free the edges of the cartilage on the sides by cutting them away from the body.

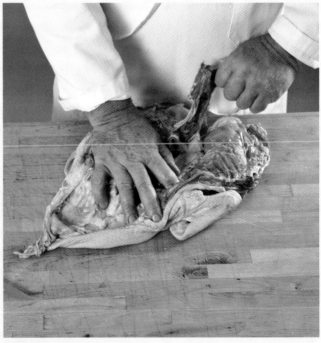

6 Remove the cartilage either by hand or using a knife, pulling it away from the body. (The lower portion of the breastbone is cartilaginous in younger birds.)

7 Cut the duck in half lengthwise through the center.

8 Cut away the excess neck skin from the head end. The skin is quite fatty and can be partially frozen, chopped finely, then simmered slowly to render its fat.

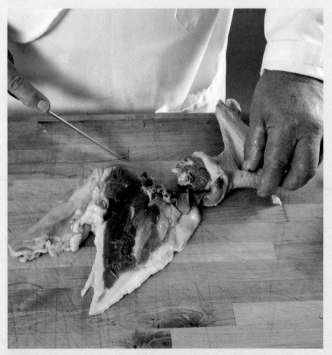

11 Wiggle the upper wing joint to find the place where it joins the breast and cut away the bony wings on both sides.

12 Trim off excess skin and fat from the breasts.

9 Slip a boning knife under the inner edge of the rib cage and cut it away, repeating on the other duck half.

10 Cut off the thigh-leg portions from either half, cutting through the skin between the breast and the thigh.

13 Turn skin-side up and score in diamond shapes diagonally through the skin but not into the meat to help the fat melt away while cooking the breasts.

14 Completed duck: At the top left are the two thigh-leg portions, next are the wings, and the carcass. At the bottom are the two scored boneless breasts, the innards, and the fat.

CONFIT

Confit, a traditional specialty of Gascony, France, is a way of preserving tougher cuts of meats, especially legs and gizzards, in usually duck, goose, and pork but also rabbit. Duck confit is often prepared from the legs of large hybrid Moulard ducks raised for their fattened livers (foie gras, page 129) and magret (breasts). The meat is salted with spices and garlic, then simmered in its own fat, covered with clarified fat, and ripened for several weeks. The confit will keep three to four months refrigerated. In earlier times, meat for confit was more heavily salted and would be stored in a cool larder up to six months.

In North Africa, lamb is similarly cured as *qadīd*, or *kedide*. The lamb is cut into thin strips, rubbed with garlic and salt, and dried. Next, it is rubbed with a hot red pepper, spearmint, caraway, and coriander seeds then sun dried like jerky. Finally, it is simmered briefly in hot olive oil and stored in the same oil.

MAKING DUCK CONFIT

In this technique, we cure smaller Pekin duck legs for confit with salt, garlic, fresh ginger, and spices. Use the same method to cure larger Moulard duck legs, increasing the amount of salt to ½ pound (225 g) and cure 3 days. Simmer the Moulard legs about 3 hours or until tender.

SEASONING MIXTURE FOR DUCK CONFIT

6 ounces (170 g) kosher salt

¼ cup (24 g) chopped fresh ginger

2 tablespoons (12 g) crushed fennel seed

2 tablespoons (13 g) crushed juniper berries

2 tablespoons (4 g) crushed bay leaves

2 tablespoons (13 g) cracked black peppercorns

1 tablespoon (3 g) dried thyme leaves

MATERIALS NEEDED:

Seasoning mixture for confit

Medium bowl for mixing spices

¹/₂ cup (70 g) peeled garlic cloves

5 pounds (2.3 kg) duck legs

1 quart (946 ml) clarified duck fat

Tray or container for curing duck

Large pot for cooking duck

Cook's fork or tongs

Pot for clarifying fat

Sieve

Large lidded container

Shown here are the seasonings for duck confit: kosher salt, fresh ginger, fennel seed, juniper berries, dried thyme, crushed bay leaf, cracked black peppercorns, and garlic cloves.

2 Place the seasoning mixture in a medium bowl and rub each duck leg entirely with it.

Arrange duck legs on a tray or in a container for curing. Cover and refrigerate 48 hours, turning the legs once. Liquid will leach out of the legs as the salt penetrates the flesh. Preheat the oven to 275°F (140°C, or gas mark 1). Rinse the duck legs well under cold water to remove salt and most of the spices then pat dry. Place in a large roasting pan (or use two pans), add the duck fat and garlic cloves, and cover with a lid or foil. Bake 2 hours or until the leg bone jiggles easily at the joint. Remove pan from the oven and allow the duck legs to cool in the fat. Using a cook's fork or tongs, remove the duck legs from the pan and transfer to a large storage container. Strain the fat through a sieve into a large pot. Simmer, skimming off any impurities, until only the clear golden fat is left, about 15 minutes.

4 Cool the fat to warm room temperature, then ladle or pour over the legs, covering them completely. Push the legs down into the fat if necessary and weigh them down with a plate. (If you don't have quite enough fat, supplement with vegetable oil.) Cool and then refrigerate to cure for 2 weeks before cooking. To cook, remove duck legs from the fat, wiping off the excess. Pan fry or roast in a 400°F (200°C, or gas mark 6) oven until the outside is crisp, pouring off any excess fat. Restaurants often deep fry the legs, a quick way to get them crispy. Use the fat for frying potatoes.

VALENTINA SANTANICCHIO:
CHEF AND OWNER OF AL SALTAPICCHIO, ORVIETO, ITALY

Valentina Santanicchio was born and raised on an organic farm, where she learned early the importance of fresh, local, and sustainable products. Located in the "Green Heart of Italy," this region of Umbria is the capital of the *Città Slow* [Slow City] movement. Returning to Orvieto after years of living in Florence, Santanicchio's appreciation of the deep traditions of food and wine that surrounded her as a girl led her to take a position at a small café in the medieval center of town. She fell in love with cooking and the restaurant world and in 2009, at age twenty-eight, she opened *Ristorante al Saltapicchio* ["the grasshopper"], where she serves innovative interpretations of classic, authentic dishes based on local, seasonal foods grown on her family's farm in a modern ambiance.

DO YOU EAT MEAT EVERY DAY?
One thing that I try never to forget is that we eat for necessity. Basically, we cannot help but kill, but we must not exceed our needs. We must avoid eating meat every day, and above all, we must make sure that the animals that we are feeding are treated with respect and that they live a life of dignity. On our farm, above all, the animals are treated with love even though they may end up on our plate, and there is always a tear for those who leave us.

WHAT DID YOU LEARN ABOUT MEAT BY GROWING UP ON A FARM?
Life in the countryside teaches many things. Above all, it teaches you that there is a natural order of things. This has always worked, but it is now in danger because people have stopped respecting the Earth and the animals. By endangering the Earth and the animals, we are putting ourselves at peril!

TELL ME ABOUT YOUR FAMILY'S FARM AND ITS HISTORY.

My family's farm is located in Ficulle, a small medieval village about 20 km [12.4 miles] from Orvieto. My paternal grandparents were sharecroppers, peasants who worked lands of others. When my father married my mother, they decided to continue that kind of work but to purchase land for themselves. So, at the beginning of the 1980s, they bought the land that they had already worked for twenty years. My father continued his work [he is a mason] and my mother and my grandparents have been working on the farm since.

WHERE IS IT, HOW LARGE IS IT, AND WHAT KINDS OF ANIMALS DO YOU RAISE?

The farm is about 20 hectares [49 acres], with diverse crops [grapes, olives, wood, and grain]. There is a large house built in the early 1900s and several stables for animals. We raise chickens, rabbits, sheep, pigs, dairy cows, and hunting dogs. Everything that my family does on the farm is in the traditional way, not intensive farming and without the use of herbicides and artificial fertilizers.

TELL ME ABOUT RAISING CHIANINA BEEF CATTLE.

The giant Chianina cattle, which is the most prevalent breed in Umbria and Tuscany, are white cows, good and calm. At birth, they are a pale pink color. On our farm, we have some Chianina cattle but also crosses of other breeds. Like our other animals, the cows spend the spring and summer outdoor grazing on pasture though they must be put in the barn during the winter. Usually one cow births a single calf although it can happen that two are born. The mother usually has only enough milk for one [Chianina cows give less milk than other breeds].

WHAT DO THE CATTLE EAT AND HOW DO THEY LIVE (ON PASTURE, FREE RANGING, IN ENCLOSED PENS)?

Our cattle graze in the fields in summer. During the times when they are feeding their calves, they eat hay that we produce ourselves on the farm and our own corn. In winter, stocks of hay and corn serve as animal feed that are forced to stay in the barn due to cold weather. In pruning season, they eat the leaves of the olive trees.

WHERE IS THE SLAUGHTER DONE—ON THE FARM OR ELSEWHERE?

The calves are transported to the slaughterhouse of Orvieto. There, after the slaughter, they are divided into anatomical parts and left for one to two weeks refrigerated for the aging then delivered directly to customers. The farm may sell directly but to no more than four customers, so only in quarters.

WHAT ARE SOME OF YOUR FAVORITE WAYS OF PREPARING BEEF?

I especially like stews, and I love to cook the "less noble" parts, such as muscolo [shank] and cheeks. I cook them for many hours with aromatics like cocoa, cloves, and spices, mainly in winter. In summer I love seared fillet with sesame or with summer truffles, and beef tartare.

WHAT ARE YOUR CUSTOMERS' FAVORITE MEAT DISHES?

In winter, slow-cooked beef with cocoa and grilled polenta; in summer, sliced beef with balsamic vinegar.

WHAT KINDS OF INNARDS DO YOU SERVE?

For veal and beef, we mostly use the liver and tongue. Sometimes, I cook them myself. For example, occasionally I make paté with onion compote and brioche bread—*delizioso!*

WHAT ANIMALS DO YOU SERVE FROM YOUR FAMILY'S FARM?

From our farm, we use rabbit, poultry, lamb, and veal. The chickens are free ranging, laying their eggs and hatching them all naturally with just a little supervision on our part. We have laying hens and breeding hens. The chickens are usually killed at one year when their meat is very tender, which we usually roast. Chickens live their lives on the ground, so their meat is very flavorful due to their diet. The rabbits are caged [otherwise they escape!]

DO YOU SERVE ANY GAME BIRDS OR ANIMALS (CACCIAGIONE), AND IF SO, WHICH KINDS?

In Umbria, you cannot buy game directly from the hunter. You must buy imported game that is very expensive and not of excellent quality.

IS RABBIT POPULAR? HOW ABOUT LEPRE (HARE)—DO YOU SERVE IT?

Rabbit, which is a courtyard animal, is very popular in Umbria. There are many traditional recipes that include rabbit. I cook it often. I have marinated, fried rabbit on my menu now—*buonissimo!* Hare is mostly a winter animal, which we use to make the best pasta sauce [famosissime le pappardelle alla lepre], the most famous pappardella [wide-ribbon pasta] with hare sauce.

ARE THERE ANY UMBRIAN MEAT SPECIALTIES THAT YOU SERVE, SUCH AS *CORATELLA*?

In Umbria, there are many dishes made with offal or with the "parts of the poor," such as tongue, liver, pluck [lungs], black pudding [sausage made with pig's blood], coppa made with ears and tail, the tail itself made with gravy and celery. There was once [it is now illegal to cook] the *pajata*—intestines of calves less than one month old containing the milk they had drunk. I like to cook traditional dishes because they remind us of our roots in simple people who knew how to adapt. In my opinion, is very important to remember the love for our land.

VENISON

Venison gets its name from the Latin *venari*, meaning to hunt, and usually refers to the meat of deer, the most common large, antlered game animal, though moose, elk, caribou, antelope, and pronghorn are also venison. Deer is farmed and ranched extensively, especially in New Zealand, where it is known as Cervena, a trade name. New Zealand is the most important source worldwide for farm-raised venison, much of it exported to the United States. This venison is fewer than three years old and comes from animals that are free to graze.

Farmed venison will be milder in flavor and more tender than wild. Ranched venison will be closer in flavor, fat content, and texture to wild deer. Venison meat is firm, moist, quite lean, and dark purplish red in color with fine grain. Wild deer may taste of forest berries, juniper, and other local plants that they feed on. Venison is widely available in European supermarkets in its traditional fall hunting season. In North America, venison and other game are usually sold frozen and may be ordered from specialty stores.

VENISON SAUSAGE WITH JUNIPER BERRIES AND PANCETTA

In this technique, we make fresh sausage in natural pork casings from a combination of leg of venison, pork butt (a shoulder cut), and pancetta (air-dried pork belly). Because venison is quite lean, mild pork fatback is added to make the ideal sausage ratio of two parts lean meat to one part fat. The sausage is flavored with shallots, garlic, juniper berries, and dried porcini mushrooms. Start a day ahead to marinate the sausage ingredients overnight in the refrigerator.

To grind the meat, we use a stainless steel hand-operated grinder (see Resources). The grinder attachment for an electric stand mixer also works well. (A food processor will yield mushy meat and is not recommended.) Stuffing into casings compresses the meat while amalgamating the flavors and makes the sausage easier to portion and to store, but the mixture may also be formed into a loaf, partially frozen, and sliced for country-style sausage.

SAUSAGE INGREDIENTS:

1½ pounds (680 g) venison leg, trimmed and chilled

1 pound (454 g) pork butt, trimmed and chilled

¾ pound (340 g) pork fatback, chilled

½ pound (225 g) pancetta

½ cup (75 g) minced garlic

½ cup (80 g) minced shallots

½ ounce (14 g, or about ½ cup) dried porcini mushrooms, soaked in warm water to cover for 30 minutes

1 tablespoon (6 g) crushed juniper berries

2 tablespoons (36 g) kosher salt

2 teaspoons (13 g) freshly ground black pepper

1 tablespoon (2 g) finely chopped rosemary

½ cup (120 ml) dry red wine

2 tablespoons (30 ml) brandy

Pork sausage casings, thoroughly rinsed if packed in salt or brine

MATERIALS NEEDED:

Scimitar or chef's knife

Scale

Meat grinder

Work surface to clamp the grinder

Coarse grinding plate

Fine grinding plate

Sausage stuffer attachment with pusher

2 large bowls

Metal tray lined with plastic wrap

Small skillet

Wooden pasta rack or cooling rack placed on a tray to catch drips

Vacuum sealer and bags, plastic wrap, or zipper-lock bags for storage

Paper towels

1 Shown here are the two trimmed and denuded leg muscles from farm-raised venison. Venison is extremely lean, so extra pork fat and pancetta are added to the sausage mixture for moisture.

2 Cut the venison into long, narrow strips and then cut the strips into small cubes. Do the same with the pork butt. Slice off the rind (hard skin) of the fatback and cut it into small cubes.

3 Shown here are the sausage ingredients: dried porcini mushrooms soaking in water to rehydrate, pancetta, venison, fatback, pork butt, salt, pepper, garlic, shallots, juniper berries, rosemary, and red wine mixed with brandy.

4 In a large bowl, combine the cubed meats with the remaining ingredients, cover and marinate overnight, refrigerated. The next day, spread the mixture out evenly on a metal tray lined with wax or parchment paper and freeze 1 hour or until the meats are stiff but not fully frozen. If working on a hot day, freeze the grinder and bowl as well so everything stays very cold. Soak the sausage casings in a large bowl of lukewarm water 30 minutes to rinse off salt or brine and to help separate the tangled hank. Flush the inside of each casing under cold running water to help open them up. Unused casings can be drained, covered with kosher salt, and frozen. Set up a meat grinder with a coarse grinding plate and attach it to the work surface with its clamp. Or, set up a standing mixer with a grinder attachment.

7 Push the sausage mixture through the grinder so that it fills the casing while incorporating as little air as possible. It is helpful to have another person help support the sausage in its casing as it comes out of the grinder. Some people prick the sausage to release any air pockets that might burst in cooking; others prefer to leave the sausage whole so the fat doesn't leach out in cooking. It's a matter of personal choice.

8 To make individual links, twist the casing every 3 inches (7.5 cm), clockwise for one twist and counterclockwise for the next.

5 Grind the mixture, ideally, first using a coarser plate using the pusher. If a finer texture is desired, chill the mixture 1 hour in the freezer and grind again using a smaller plate. To test for seasoning (never eat raw pork), flatten a walnut-size clump of the sausage mixture. Cook in a small skillet over medium-low heat until browned and cooked through before tasting. Adjust seasonings as necessary.

6 Assemble a sausage stuffer, here, a plastic tube placed inside the grinder without the grinding plate. Load a length of natural pork sausage casing onto the tube, carefully untwisting the casing while pushing it onto the tube. Avoid tearing the casing and leave the end untied.

9 Tie a short length of butcher's string between each pair of links, cutting off the excess string with scissors. Allow the sausages to dry at cool room temperature for 1 hour hung on a pasta rack or placed on a wire cooking rack for good air circulation. Place the sausages in a container lined with paper towels to absorb any drips. Refrigerate up to 2 days before cooking, or vacuum seal or place in plastic zipper-lock bags and freeze.

10 Completed venison sausage links.

Photo: © Holly Heiser, 2011

HANK SHAW:
GAME EXPERT AND AUTHOR OF *HUNT, GATHER, COOK*, SACRAMENTO, CALIFORNIA

Hank Shaw, best known for his blog Hunter, Angler, Gardener, Cook, is a hunting enthusiast who writes about the food that he fishes, hunts, forages, and gardens. A former line cook, newspaper reporter, and commercial fisherman, Shaw is the author of *Hunt, Gather, Cook*, which was published by Rodale Press in 2011. A forager and angler since childhood, Shaw began hunting in 2002 and has never looked back. Shaw and others who share his viewpoint are part of a new wave of eco-conscious hunters who hunt to feed themselves from the land, to connect with the natural cycle, and to eat meats replete with flavors of the regions they inhabit, while making connections with strong hunting traditions.

HOW AND WHY DID YOU START BLOGGING ABOUT WILD FOODS?
I was the capital bureau chief for the *Stockton Record* in Sacramento where I blogged successfully about politics. Once I got to see the power of blogging firsthand, I started my own blog in 2007 about hunting, fishing, and foraging, which at the time was my avocation.

DID YOU COME FROM A HUNTING FAMILY?
No, I came from a foraging and fishing family in north Jersey. No one that I knew growing up was a hunter. In 1994, a couple years after I first started as a reporter on Long Island, I managed to collect, catch, and grab fish, shellfish, and crabs—every bit of protein that I ate from March until Christmas. Years later, a friend suggested that I take up hunting and took me on my first hunt in South Dakota. I had never fired a shotgun before. We practiced on milk jugs. At first, I couldn't hit the broadside of a barn, but my friend taught me. Though I had a lot of fishing experience, hunting on land takes a different set of skills and knowledge. I got my license and began with rabbits and squirrels, pigeons and doves.

WHEN AND WHERE DO YOU HUNT FOR GAME?

I hunt wherever and whenever. I've done most everything, but I don't shoot what I don't eat. I try to hunt at least one big game animal, like deer or wild boar, every year though I've never hunted bear. I also spend a lot of time hunting birds.

WHAT IS THE HUNTING YEAR?

There is a rhythm to the year. The traditional hunting year starts on Labor Day with doves. Deer, however, may start in August, depending on where you are; our season lasts till November. In October and November, we hunt traditional upland birds like turkey, quail, pheasant, grouse, and partridge. We also get jackrabbits, cottontails, and snowshoe hares in the Sierras. Winter we spend hunting ducks and geese until the season shuts down at the end of January. Spring starts with turkey, which also has a season before Thanksgiving. We hunt feral hogs all year.

IS HUNTING A FAIR CONTEST?

One of the things about hunting is that you have to be there to "get it." It may not look like it, but hunting is a meeting of equals. When I'm hunting, I'm not going to necessarily succeed. All these animals I'm hunting are at their full capacity as smart, swift, and stealthy creatures. Some days, it's easy, some days we catch nothing. I also think that if the roles were reversed, I'd much rather be hunted than raised as livestock. For a domestic animal, I raise it. A wild animal raises itself. Hunters have a lot more respect for what we chase than the average person has for a pig or a cow.

WHAT IS "THE FORGOTTEN FEAST"?

I came up with that phrase to remind people that regular Americans once ate game commonly, and to honor traditional preparation skills like butchering, curing, and smoking. We have many forgotten foods: Muskrat was sold in the finest restaurants in the U.S. Delmonico's in Manhattan sold it as "marsh hare," and there was no better restaurant in the country at the time.

DO YOU FAVOR EATING FARMED GAME?

I think we should be eating more farmed venison and bison. The Great Plains were designed for bison. If ranchers restore the land to native grasses, bison won't wreck it. And bison meat is so similar to grass-fed beef that it takes an expert to tell the difference. If I were to buy meat, I would buy bison, which, by nature, is raised free range. I vehemently oppose factory farming—I don't want to eat beakless chickens or docked pigs [with their tails removed]—so when I eat out, I often go vegetarian.

WHAT DO YOU RECOMMEND FOR THE NOVICE HUNTER?

Start small with rabbits and squirrels, where the kill is less momentous, then move on to bigger game. When you hunt small game you need to be stealthy, be quick, and understand habitat. Those skills translate to other animals. Plus, there are way more opportunities to hunt small game. And with a deer hunt, you may pull the trigger once. You might do it four to ten times a day when hunting rabbit.

DO YOU NEED TO BE IN A SPECIAL STATE OF MIND WHILE HUNTING?

You must have the ability to quiet your mind, be patient, and let everything around seep into you. The biggest issue is being fidgety. I can sit still for three hours and become part of the surroundings. Every time you move, it's like throwing a rock in a pond. A good hunter lets those ripples fade before moving again. Animals need to forget you're there. Learn to look with your eyes and move your head very slowly.

HOW DO YOU PRESERVE THE MEAT?

Mostly, I vacuum-seal and freeze the meat. Because I eat everything on the animal, I do a lot of salami, some smoking, and a lot of sausage. I especially like to make salami, lonzino, and Cajun tasso ham from wild pigs.

HOW ARE THOSE INTERESTED IN SUSTAINABILITY AND TRADITIONAL HUNTERS SIMILAR AND DIFFERENT?

It is an interesting mix. One of my goals is to get the two groups to start talking to each other because they have common interests. There are a lot of people for whom game is their primary source of protein, mostly rural folks in the South, the mountain West, and Alaska. But, and this is especially true in California, many food people also want to start hunting. Once they get their license and learn to shoot, I will often take them out to show them the ropes.

WHAT TRADITIONAL HUNTER'S STEWS DO YOU MAKE?

Italian cacciatore, Polish bigos, American Brunswick stew, French chasseur, Hungarian goulash, Southwest American chili con carne, and Spanish chilindron. Pretty much every culture has its version of a hunter's stew. I love them all.

RABBIT

The domestic rabbit has mild-tasting lean pink meat and is closely related to its wild cousin, the hare, which has stronger-tasting dark purplish red meat. Rabbit is popular in Europe, especially in France, Spain, Italy, and Germany, parts of the Middle East, and South America, less so in North America (which is home to half the world's rabbit population). A rabbit fryer is, like a chicken fryer, a young animal with finely grained, tender flesh, and an average weight of 2½ pounds (1.1 kg). A roaster rabbit, which weighs about 5 pounds (2.3 kg), has firmer flesh with more prominent grain and stronger flavor. Heritage breed rabbits including American Chinchilla Silver Fox are available in the United States from specialty farms.

When handling wild rabbit, it is a good idea to use gloves as it may carry tularemia, or rabbit fever. While it is possible for domestic rabbits to be infected, instances are much rarer. Rabbits are commonly hunted in the wild but for commercial sale, they must be farm raised. Because its meat is so lean, rabbit has a tendency to become dry if cooked at too high heat or over-cooked. Marinate rabbit and cook at low heat to maintain moisture. For mature rabbit, soak in saltwater before cooking for milder flavor. The wild hare is a close cousin to the rabbit, which may be wild or domestic, and has similar body structure but leaner, darker, stronger-tasting meat. Other names for rabbit include *Kaninchen* (German), *conejo* (Spanish), *coney* (old English, source of Coney Island), *coniglio* (Italian), and *lapin* (French).

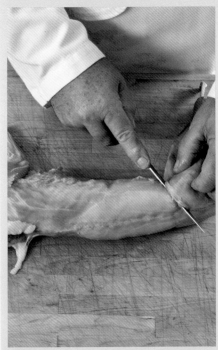

1 Cut through the flesh just forward of the rear leg starting on the outer side until you reach the ball-and-socket joint connecting the leg to the pelvis.

FABRICATING WHOLE RABBIT

Fabricating means breaking down primal meat cuts into usable parts. Here, we fabricate a whole (headless) rabbit, a sustainable meat due to its legendary mating habits. The saddle of rabbit (rib and loin) is most tender and juicy; the meaty rear legs represent about 40 percent of the dressed rabbit's weight with more pronounced muscle grain; the front legs contain less meat and are stringier with fragile bones.

MATERIALS NEEDED:

Boning knife

Meat cleaver or heavy chef's knife

Heavy kitchen shears (optional)

Vacuum sealer and bags, plastic wrap, or zipper-lock bags for storage

4 Turn the rabbit around, grasp the body, and keeping your nondominant hand safely away, use a cleaver (or heavy chef's knife) to chop off the pelvis bones. The neck and pelvis are tough parts with good flavor but little meat, best used in the stockpot.

2 Grasp each leg and twist to pop them from their joints. Remove the legs on either side of the rabbit. Lift up the front legs and sever from the body, cutting through the tendons connecting the upper legs to the shoulder.

3 Place the rabbit torso on the work surface with ribs facing up. Grasping the rear and keeping your nondominant hand safely away, use a cleaver (or heavy chef's knife) to chop off the neck bones.

5 Split the rabbit body by cutting through the rib cage, holding the body in place with your other hand.

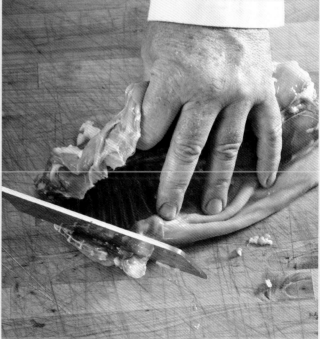

6 Pull open the rib cage section and chop off the rib bone ends on either side using a cleaver, or cut them away with heavy kitchen shears.

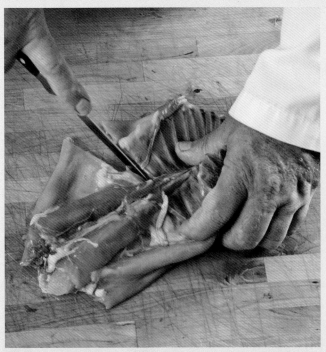

7 Cut through the center of the body but not the spine bone, between the end of the ribs and the loin section on either side of the body.

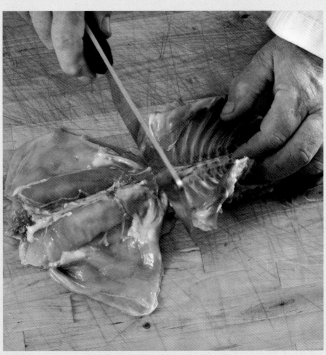

8 Using a cleaver or a heavy chef's knife and keeping your nondominant hand safely away, chop through the spine bone, cutting into rib and loin sections.

9 If desired, remove the loin and attached belly from the bones by cutting underneath the muscle starting at the spine and cutting outward. The boneless loin may be stuffed, rolled, and tied before cooking.

10 Re-formed rabbit broken down into its main usable parts: rib section, loin section, front and rear legs.

Though much leaner, goat has similar bone structure to lamb and may be prepared in the same way.

ABOUT LAMB

Lamb is a member of the Bovidae family along with cattle, goat, and bison, all ruminants with a four-chambered stomach that allows them to digest plants as they graze. First domesticated about 10,000 years ago in Central Asia, sheep herd easily, can tolerate more severe conditions than cattle or pigs, and provide meat, wool, and rich milk. Like the meat from its cousin, the goat, lamb is distinctively earthy, even gamy in flavor, and the more mature the lamb, the stronger its flavor and darker its color. Lamb is important in the cuisines of the Mediterranean region, Turkey, central Asia, the Indian subcontinent, northern China, and Indonesia as well as Scotland and Wales in the UK and Australia and New Zealand, which are both important exporters. Though much leaner, goat has similar bone structure to lamb and may be prepared in the same way.

Lamb reached the United States with the Spanish explorers, but Americans eat far less lamb than beef, pork, or poultry—about 1 pound (454 g) per person per year—partly because its introduction into Western cattle herds in the nineteenth century led to fierce battles between shepherds and ranchers. Basque immigrants who began as shepherds in the nineteenth century dominate the U.S. Western lamb industry. China has the largest sheep herd in the world, but New Zealand is the biggest exporter, much of it to Europe and the UK where smaller animals are preferred. Until after World War II, mutton was favored in the UK, when tastes changed to prefer milder, younger, lamb though mutton is making a comeback especially on restaurant menus. Elsewhere, mutton is enjoyed in great quantity in France, the Caribbean, Africa, the Middle East, India, parts of China, Australia, and New Zealand.

Because the structure of the lamb rib (or rack) is exactly the same as for the veal rack, pork rib section, and venison rack, differing only in size, shape, and color, follow the instructions on page 64 to prepare rack of lamb, pork, or venison. Other names include *agneau* (lamb) and *mouton* (mutton) in French; *agnello* (lamb) and *moutone* (mutton) in Italian; *cordero* (lamb) and *ovino* (mutton) in Spanish; and *Junglamm* (young lamb), *Schaffleisch* (lamb), and *Hammelfleisch* (mutton) in German.

AVERAGE SIZE OF LAMB

Grain-finished American lamb is larger and milder in flavor than Australian and New Zealand lamb, which feed on grasses.

New Zealand	33 pounds (15 kg)
Australian	40 pounds (18 kg)
American	65 pounds (30 kg)

TERMS FOR LAMB (DEFINITIONS VARY FROM COUNTRY TO COUNTRY)

Milk-fed lamb: Meat from an unweaned lamb, four to six weeks old, found in Europe but not the United States and the UK, because of its high cost. Known as *abbacchio* in Italy.

Lamb: The meat of a young sheep under one year of age.

Saltbush mutton: Australian term for the meat of sheep that graze on saltbush plants.

Agneau de pré-salé (salt marsh lamb): French term for the meat of sheep that graze on salt marshes and also found in the UK.

Hogget: The bolder-flavored meat of a young adult lamb, older than lamb and younger than mutton.

Mutton: The strong-tasting meat of sheep over two years of age, which may be dry-aged like beef. The same term refers to goat meat (or chevon) in India, Malaysia, and Singapore.

MATERIALS NEEDED:

Boning knife

Scimitar knife

Honing steel

Vacuum sealer and bags, plastic wrap, or zipper-lock bags for storage

DEBONING LEG OF LAMB

In the United States, lamb is cut straight across the carcass at the packing house and divided into five primal cuts: breast and foreshank, chuck, rib, loin, and leg.

Here, we debone a domestic American leg of lamb (NAMP 233A) weighing about 10 pounds (4.5 kg).

1 Shown here is the inner side of a bone-in leg of lamb. The small, detached trotter bone is separated for easier shipment but left attached by the large tendon. If present, pull off the thin, papery membrane, called the fell, which covers the outside of the lamb leg.

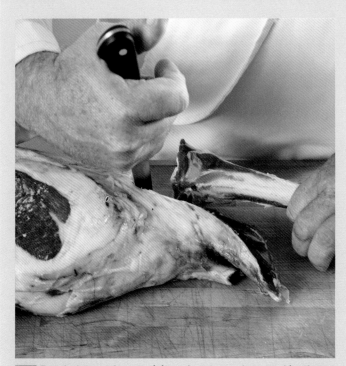

4 Turn the bone at the upper joint and cut any tendons attaching the two bones, then cut away and release the shank bone.

2 Find the joint between the lower shank bone and the trotter and cut between them, removing the bottom portion of the bone with its attached Achilles tendon.

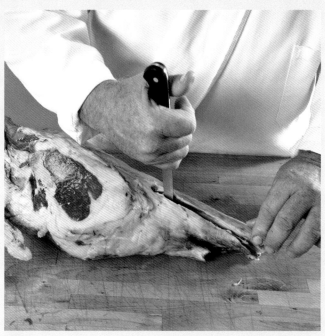

3 Cut along the outside of the shank bone below the stifle (knee) joint, then cut along the inside up to the joint with the large leg bone to expose the bone.

5 Remove the flaps of flank and top sirloin that cover the hip bone. (Alternatively, remove the tough white membrane from between the flank and top sirloin then remove the covering from the outside, leaving the meat attached to add more weight to the leg).

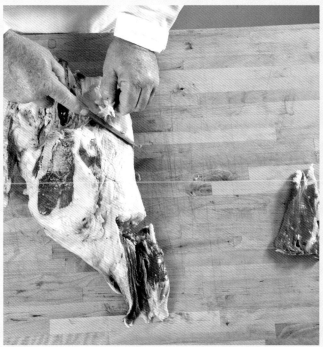

6 Trim off the heavy fat from the inner side of the leg.

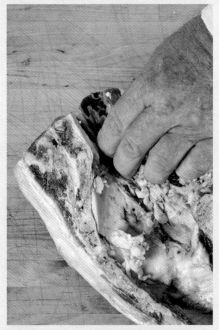

7 Find the tube-shaped tenderloin, insert the tip of the knife under the tenderloin and over the pelvic bone, and cut along the curved outer side of the pelvic bone.

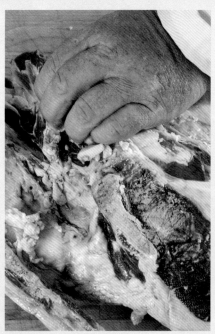

8 Cut under the tenderloin and around in a curve following the underlying pelvic bone, which curved to meet the ball-and-socket joint at the upper end of the large leg bone.

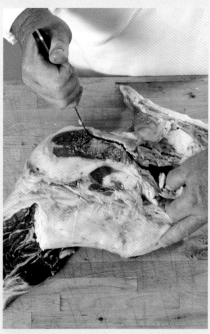

9 Cut around the edges of the pelvic bone, following its curved shape up and over the rounded ball-and-socket joint and continue past it until you reach the spine.

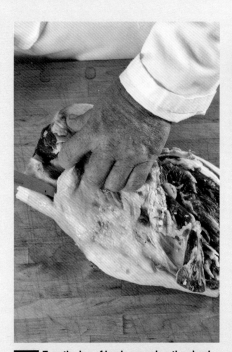

13 Turn the leg of lamb around so the shank end faces your knife and cut as close as possible around the femur bone in tunnel form. (Alternatively, slice the leg open along the side of the femur bone and then remove—an easier but less elegant way to bone the leg.)

14 Turn the leg around so the hip end faces your knife and cut as close as possible around the femur bone, tunneling it out to remove it.

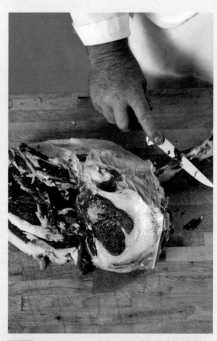

15 Pull the femur bone out from the bottom while scraping away any remaining meat with your boning knife. Remove the cartilaginous patella (kneecap) below the bottom end of the femur bone.

10 Pulling the large pelvic bone away from the leg bone using your nondominant hand, insert the tip of the knife under the pelvic bone to cut the tendons connecting the leg bone to the pelvic bone.

11 Pull off the pelvic bone, exposing the top rounded ball socket of the large femur bone.

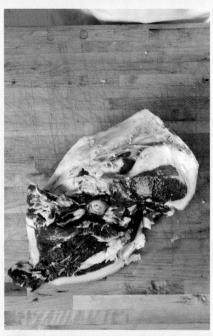

12 Cut away the shank meat below the stifle (knee) joint. Leg of lamb with shank meat to the side.

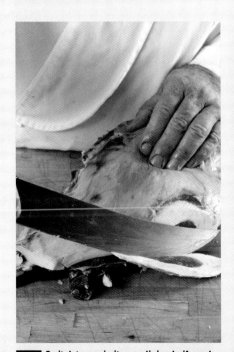

16 Switch to a scimitar or slicing knife and cut away most of the heavy layer of fat from the outside of the now boneless leg.

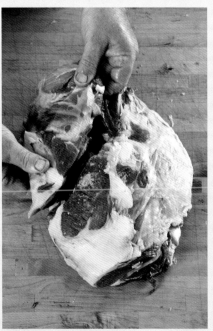

17 As shown, place the reserved shank meat inside the opening of the lamb leg where the pelvic bone was located to make a roast of even thickness.

18 Tie the boneless leg of lamb with butcher's string (page 12), starting with the first string in the center and then adding two more at either end of the leg.

ABOUT CAUL FAT

A chef's favorite for French charcuterie (cooked meats) and to moisten small birds as they roast, caul fat is the transparent, fragile, lacy lining of the stomach of a pig or lamb, though pig caul is preferred. To use, defrost and soak the caul fat in warm water until pliable, then drain. Unused caul fat may be refrozen. As the caul fat melts away in cooking, it moistens the meat, helps keep it in a compact shape, and creates a brown crust.

In France, where caul fat is especially valued, crépinettes are small, flat packets of sausage meat enclosed in caul fat. Caul fat is usually sold frozen and may be available at Asian, French, and Italian markets, or by special order. Other names for caul fat include: *crépine* (French), *Bauchnetz* (belly net, German), *rete di pancia* (Italian), *lace fat* (English), and *redaño* (Spanish).

CYPRIOT SHEFTALIA SAUSAGE PATTIES

Sheftalia are Cypriot lamb and sausage patties that are wrapped in caul fat. In Atlantic France, similar crépinettes, which get their name from the French for caul fat, accompany oysters on the half shell. The sheftalia here are made of ground pork and lamb mixed with chopped onion and parsley and formed into patties, which are

placed on squares of spread-out caul fat and wrapped. The sheftalia may be skewered and grilled over charcoal in the traditional way or they may be pan-seared until well browned and cooked through. Sheftalia freeze well up to 2 months if vacuum sealed; otherwise, store refrigerated 3 to 4 days or freeze for 2 to 3 weeks.

MAKES ABOUT 30 SHEFTALIA:

1 pound (454 g) pork shoulder, trimmed

1 pound (454 g) lamb shoulder, trimmed

¼ pound (113 g) pork fatback

1 large onion, finely chopped

½ cup (30 g) chopped flat-leaf parsley

3 tablespoons (18 g) chopped spearmint or 1 tablespoon (2 g) dried spearmint (nana)

2 teaspoons (5 g) ground cinnamon

2 teaspoons (12 g) salt

Freshly ground black pepper to taste

½ pound (225 g) pork caul fat, defrosted if frozen

Cut the pork and lamb into 1-inch (2.5 cm) cubes. Combine the meat in a metal bowl or on a metal tray and freeze until stiff but not fully frozen, about 1 hour. Remove the rind, if present, from the fatback and freeze until stiff but not fully frozen, about 1 hour. Chill the fatback, then cut into very small cubes. Set up a meat grinder with a coarse grinding plate, attaching it to the work surface with its clamp. Or, set up a standing mixer with a grinder attachment.

MATERIALS NEEDED:

Meat grinder
Coarse grinding plate
Scale
Work surface to clamp the grinder
Fine die
Pusher
Small skillet
Sausage stuffer
2 bowls
Scissors
Vacuum sealer and bags, plastic wrap, or zipper-lock bags

1 Using the pusher, grind the meat mixture. Combine the ground meat with the remaining ingredients except caul fat in a bowl, kneading well with your very clean or gloved hands until pasty. To test for seasoning (never eat raw pork), flatten a walnut-size clump of the sheftalia mixture. Cook in a small skillet over medium-low heat until browned and cooked through before tasting. Adjust seasonings as necessary.

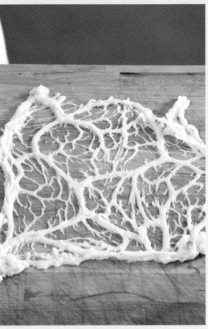

2 Place the caul fat in a bowl of warm water, which will soften it. Carefully unravel the fragile caul fat and stretch it out over your work surface. (Don't worry if the caul fat tears; it's okay to piece it together.)

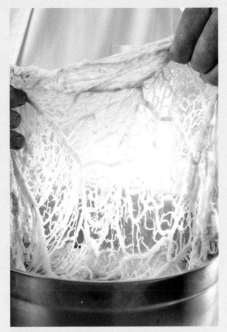

3 Cut the caul fat into 4-inch (10 cm) squares and arrange on the work surface.

4 Use your hands to form the sheftalia mixture into compact patties, 2 to 3 ounces (55 to 85 g) each. Place one patty in the center of each caul fat square.

5 Wrap the caul fat so that it completely encloses the meat, trimming off excess caul fat with scissors.

6 Completed sheftalia. Allow the sheftalia to rest in the refrigerator for 4 hours before cooking. Store by vacuum sealing and freezing for best results. To cook, defrost sheftalia and either brown on all sides in a heavy skillet or grill until well browned and the internal temperature reaches 145°F (63°C).

I respect those who raise strictly grass-fed animals, but I believe that grain is critical to flavor, so we supplement their diet with grain and soybeans.

LINDA GEREN:
SUSTAINABLE MEAT FARMER, MOUNT HOLLY, NEW JERSEY

Linda Geren and her husband, Michael McKay, humanely raise pigs, goats, sheep, and heritage breed chickens at their 20-acre [8 hectare] High View Farm near Mount Holly, New Jersey. Geren grew up on a large farm in the Ozark Mountains where her best friends were farm animals. Chefs and eco- and quality-conscious foodies in the Philadelphia, New York, and Princeton markets are happy customers for her lamb, pork, and free-range chicken eggs. Two things drive Geren: great flavor in the meat and good animal welfare.

HOW WAS THE MEAT PREPARED ON YOUR FAMILY'S FARM?
The meats, especially pork, would be butchered in fall so that it could hang in cold weather. My father would sugar-cure his hams, which were fairly salty, like country ham. We would cook small pieces with red-eye gravy made from the pan drippings. We didn't eat lamb growing up. We did have sheep for a short time, but they were kept for their wool.

WHAT ANIMALS DO YOU RAISE AT HIGH VIEW FARM?
We raise animals for meat including pigs, sheep, and now goats. We also keep chickens and board horses. Our pigs are the Landrace breed, which is of Danish origin and brought to the U.S. 250 years ago. The breed is known for its long loin and body, very good tasting meat, and high production.

WHERE ARE YOUR ANIMALS SLAUGHTERED?

We use a small local family-run butchery called Bringhurst Meats, which was founded in 1934. Two Bringhurst brothers began killing pigs to sell along with the sweet potatoes they grew on their farm. Soon other farmers began bringing their animals to be processed. Today, it is still run by the family in the old farmstead. They process and sell beef, pork, veal, lamb, goat, ostrich, emu, and buffalo under USDA inspection and make hams, bacon, pork sausage, and other processed meats.

HOW DO YOU GET THE ANIMALS TO THE SLAUGHTERHOUSE?

I keep the animals at the farm until hours before the end, so they're fed and cared for in family groups, which reduces stress. While we can't remove all the fear, we do all the hauling ourselves. It's common practice to drop off animals over the weekend, so they might spend a day or two in the holding pens. I don't do that, because the environment stressful and I'm sure it affects the flavor and texture of the meat.

YOU HAVE SUCH A CLOSE RELATIONSHIP TO YOUR ANIMALS. HOW DO YOU FEEL WHEN YOU BRING THEM TO SLAUGHTER?

It's always a hard day, but you learn to separate yourself from that aspect. I try not get terribly attached to animals designated for butchering. Sometimes you have to make difficult decisions as to which ones to keep for breeding. I've watched the whole process and have done some butchering myself, but I don't enjoy it at all.

WHY DO YOU RAISE PIGS?

It's never been an economic decision to do this, because it's a huge amount of work and very expensive to raise pigs (and lambs) to full grown. It has to do with me knowing where the meat came from and wanting to re-create the flavor and quality

of meat I grew up with. I could go to the store and pay much less for my meat with much less work, but it wouldn't taste the same and it wouldn't be the same!

ARE YOUR ANIMALS GRASS FED?

I respect those who raise strictly grass-fed animals, but I believe that grain is critical to flavor, so we supplement their diet with grain and soybeans. To my mind, we've gone overboard in the perception of grains representing something bad in the diet. In the past, animals were fed what was growing locally, so flavor differences were regional. In the Ozarks, our pigs ate acorns, because we were in the middle of a huge grove of trees. In the South, animals eating ramps [also known as wild onions or wild garlic] in spring would impart that taste to the meat.

IS HIGH VIEW FARM ORGANIC?

No, because so much is involved in becoming certified. We buy our piglets from our neighbor's farm, but to be certified organic, all the animals must be born on the farm. Every product they eat must be organic certified, an ordeal that didn't seem worth it to us. To me, it's most important that the animals be treated well, so we are Animal Welfare–approved, as is our butcher. The certification verifies that the animals are treated in the most humane way possible from birth to death.

Lamb shoulder complete with upper foreshank shown from the inside

DEBONING SHOULDER OF LAMB

MATERIALS NEEDED:

Boning knife

Vacuum sealer and bags, plastic wrap, or zipper-lock bags for storage

The full-flavored, fatty lamb shoulder (NAMP 207A) is preferred over leaner, denser lamb leg meat for grinding, stews, kabobs, and satay strips, and for boneless rolled and tied roast. The shoulder used here includes the upper foreshank and is sometimes known as a lamb oyster shoulder. A square-cut lamb shoulder (NAMP 207) will have had the foreshank removed and may include anywhere from three to seven ribs. Besides the ribs, the main bones contained in the shoulder are the blade, the upper arm bone (or humerus), and the two bones (the larger ulna and the smaller radius) inside the lower arm. Other names for lamb shoulder include *paleta* (Spanish), *épaule* (French), *spalla* (Italian), and *Schulter* (German).

1 Slide the knife underneath the meat covering the blade bone keeping the knife hugging the bone. Cut away and lift up the meat, thereby exposing the blade bone, which is spatula-shaped on top and ridged underneath. Keep the meat attached on one side.

2 The blade bone is attached at the shoulder end to the upper arm bone with a ball-and-socket joint. Cut between the two bones at the smaller end of the blade bone where it connects to the upper arm bone.

3 Cut all around the edges of the blade bone and then pull it up and out, using the knife to free it partially from the shoulder.

4 While steadying the shoulder with your dominant hand, pull the blade bone away from the shoulder, removing it completely and cutting if necessary to release it.

5 With its inner side facing up, cut down alongside the upper portion of the large upper arm bone, or humerus.

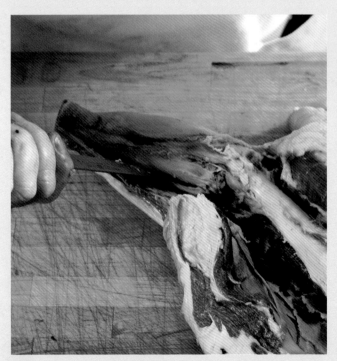

8 Cut the meat away from around the two lower arm bones.

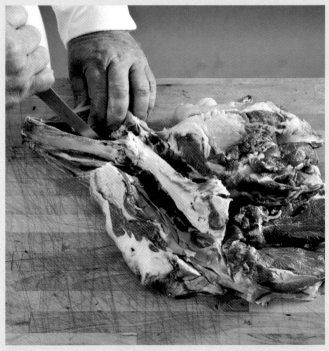

9 Cut and pull the upper and lower arm bones away from the lamb shoulder.

6 Shoulder of lamb opened up with partially exposed humerus bone visible.

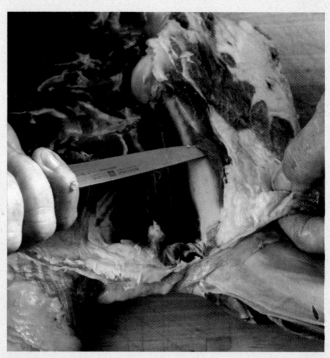

7 Pull open the flap of meat and fat covering the humerus bone, cutting around the bone to release it.

10 Remove the arm bones from the shoulder.

11 Boneless lamb shoulder shown from the inner side. Trim off excess fat, fat pockets, and silverskin before rolling and tying for roast or cutting into cubes or strips.

RESOURCES

CONTRIBUTORS

ACTIVA
www.activatg.com
Information about transglutaminase, "meat glue," produced in Japan by the Ajinomoto Corporation. To purchase, visit online retailers.

ANISSA HELOU
www.anissas.com
Website about Anissa Helou, author of many books about Mediterranean cuisine including *The Fifth Quarter*, about organ meats.

ARTISAN BEEF INSTITUTE
www.artisanbeefinstitute.com
Founded by Carrie Oliver to evaluate, rate, and create professional tasting notes for individual farms, ranches, and butcher teams.

CAPLANSKY'S DELICATESSEN
www.caplanskys.com
Toronto Jewish-style deli specializing in house-smoked, hand-cut Montreal-style smoked meat founded and owned by Zane Caplansky.

CLAGETT FARM
www.clagettfarm.org
A program of the nonprofit Chesapeake Bay Foundation, raising produce and meats using economically and environmentally sustainable farming methods and selling grass-fed beef biannually as a "split half."

CREATIVATORS
www.creativators.com
Founded by president and CEO Gene Gagliardi, this company specializes in developing unique food products from by-products and underutilized cuts.

DR. TEMPLE GRANDIN
www.grandin.com
Research articles by Dr. Grandin about livestock handling, slaughterhouse design, animal welfare, ritual slaughter, humane slaughter, and stress and meat quality.

FANTE'S KITCHENWARE
www.fantes.com
http://www.fantes.com/knives.html#boning
(page about knife characteristics and materials)
Store and website selling a large variety of kitchen tools including many lines of knives; excellent background information about purchase and care of knives and other meat-cutting tools; professional knife sharpening service.

Tools used in this book from Fante's include

5-quart (4.7 L) KitchenAid mixer with meat grinder attachment and pusher
Butcher's string
Jaccard cutter
Larding/trussing needle
Meat grinder plates
Meat pounder
Plastic sausage funnel kit
Roast wrap (netting for boneless roasts)
Tre Spade meat grinder (imported from Italy)

GEORGE L. WELLS MEAT COMPANY
www.wellsmeats.com
Founded in 1908, this Philadelphia-based, independently owned company is a purveyor of top quality, dry- and wet-aged meats, Certified Angus Beef, poultry, game, and specialty foods to top restaurants, hotels, caterers, and clubs.

GROW AND BEHOLD
www.growandbehold.com
Orthodox Union certified, Glatt Kosher pastured meats raised on small family farms and adhering to the strictest standards of *kashrut* (kosherness), animal welfare, worker treatment, and sustainable agriculture owned by Naftali Hanau.

HERITAGE MEATS
www.heritagemeatswa.com
Artisan butchering company specializing in locally grown and sustainable meat products and services. Certified USDA meat processor and charter member of the Puget Sound Meat Producers Cooperative owned by Tracy Smaciarz.

HIGHVIEW FARM
www.highview-farm.com
Farm near Mount Holly, New Jersey, producing naturally raised pork, lamb, and soon, goats. Co-owned by Linda Geren and Michael McKay.

HUNTER, ANGLER, GARDENER, COOK
www.honest-food.net
Hank Shaw's blog with recipes about wild game and humanely raised meats, foraging, gardening, and cooking.

AL SALTAPICCHIO
www.alsaltapicchio.com
Restaurant in Orvieto, Italy, with most food ingredients, including heritage Chianina beef, organic and local from Chef Valentina Santanicchio's family's farm.

KOBE BEEF AMERICA
www.kobe-beef.com
Company founded by R. L. Freeborn breeding and raising Wagyu cattle to Japanese specifications, selling Wagyu beef in the United States and exporting to Japan.

LA QUERCIA
www.laquercia.us
Herb and Kathy Eckhouse produce Italian-style artisan dry-cured meats, or salumi, including prosciutto, pancetta, guanciale, and lardo. Varietal meats, some from heritage breeds, some organically raised, all from sustainable producers.

LA TIENDA
www.tienda.com
Bilingual family company owned by Don Harris supporting artisanal and small family farms in Spain with the largest selection of Spanish food online, including pata negra ham.

SENAT POULTRY
www.enameatpacking.com/senat_poultry
Halal chicken slaughterhouse in Paterson, NJ, selling fresh, humanely raised head-on, foot-on chickens in the New York City region.

WÜSTHOF KNIVES
www.wusthof.com
Information about Wüsthof knives and butcher's tools produced in Solingen, Germany. Wüsthof tools used in this book include

10-inch (25 cm) scimitar knife

2-stage manual sharpener

Wüsthof whetstone

Classic 5-inch (12.5 cm) forged boning knife

Classic 6-inch (15 cm) cleaver

FOR MORE INFORMATION

ANIMAL HANDLING
www.animalhandling.org
American Meat Institute–sponsored website about animal handling with resources about government regulations and voluntary guidelines to promote optimum livestock handling.

ANIMAL WELFARE INSTITUTE
www.awionline.org
A nonprofit organization founded in 1951 working to improve all aspects of a farm animal's life including breeding, growing, transport, and slaughter.

ARGENTINEAN BEEF CUTS CHART
http://sweethomefloripa.com
http://sweethomefloripa.com/wp-content/uploads/2009/09/Cortes-Varias-Linguas-WEB11.pdf Includes cut names from Argentina, Brazil, Spain, France, and many Latin American countries.

AUSTRIAN MEAT MARKETING
www.fleisch-teilstuecke.at/en/98/
Website showing Spanish, French, German, Italian, English, and Austrian names of beef, veal, pork, and lamb cuts from Agrarmarkt Austria Marketing GesmbH, a service organization of Austrian agriculture.

BEEF RESEARCH
www.beefresearch.org/default.aspx
Detailed information about U.S. beef from the National Cattlemen's Beef Association including fact sheets and topic briefs.

BEEF RETAIL
www.beefretail.org
Sales data, research, recipes, nutrition, and resources about selling beef and veal at retail including. Funded by the Cattlemen's Beef Board and National Cattlemen's Beef Association.

BOVINE MYOLOGY
http://bovine.unl.edu/bovine3D/eng/
Beef anatomy from the University of Nebraska at Lincoln with 3-D muscle ID, fabrication videos, skeleton, cross-sections, lateral views, subprimals, and bone and muscle descriptions with common and scientific names.

BUTCHERS ONLINE MEAT COMPANY
www.butchersonlinemeat.co.uk/beef-cuts.html
Information and chart showing British beef cuts with cooking recommendations.

CANADIAN MEAT CUTS
www.inspection.gc.ca/english/fssa/labeti/mcmancv/mcmancve.shtml
Canadian meat cuts manual for beef, veal, lamb, poultry, and pork in English and French.

CLOVE GARDEN
www.clovegarden.com/ingred/ab_cowc.html
Information about U.S. beef cuts with photos of some cuts and links to Australian, UK, and Canadian beef cut charts.

DAVISON'S BUTCHER SUPPLIES
www.davisonsbutcher.com
Large selection of professional butcher supplies.

EFFECT OF MARBLING ON BEEF PALATABILITY
http://jas.fass.org/content/72/12/3145.full.pdf
Article from the *Journal of Animal Science* referred to in interview with Carrie Oliver: "Effect of marbling degree on beef palatability in *Bos taurus* and *Bos indicus* cattle."

EUROPEAN UNION ANIMAL HEALTH AND ANIMAL WELFARE
http://ec.europa.eu/food/animal/index_en.htm
Protecting and raising the health status and condition of animals in the European Union, in particular food-producing animals.

FRENCH CUTS OF MEAT
www.frenchentree.com/france-food-cuisine/displayarticle.asp?id=2189
Website comparing French and British cuts of meat.

ITALIAN MEAT CUTS
www.italianeating.eu/what_means/meat/beefcuts/firstclassbeefcuts.htm
Information, diagrams, and photos showing Italian beef cuts.

ITALIAN PORK CUTS
www.italianeating.eu/what_means/meat/porkcuts/porkmeatcuts.htm
Information, diagrams, and photos showing Italian pork cuts.

KOBE BEEF
www1.american.edu/ted/kobe.htm
Case study about Kobe beef from the TED Institute.

KOSHER SLAUGHTERING
www.kosherquest.org/bookhtml/_.htm
Detailed information about kosher slaughtering, preparing kosher meat, and kosher butcher shops.

MEAT CURING
www.extension.umn.edu/distribution/nutrition/DJO974.html
Information about meat curing from the University of Minnesota.

MEAT CUT CHARTS
www.virtualweberbullet.com/meatcharts.html
U.S. meat cut charts.

MEAT EDUCATION AT TEXAS A&M UNIVERSITY
http://meat.tamu.edu/ANSC307lab.html
Course material from Texas A&M University covering meat inspection, anatomy, pork slaughter, and dressing, pork evaluation, curing and smoking of meat, sausage manufacturing, lamb slaughter and dressing, lamb evaluation, lamb fabrication, beef slaughter and dressing, beef evaluation, beef fabrication, quality control and packaging procedures, and meat cookery.

PIG ON A SPIT
www.pigonaspit.com/pork-cuts.php
International reference for cuts of pork.

PORK INDUSTRY INSTITUTE
www.depts.ttu.edu/porkindustryinstitute
Documents, links, and resources related to the U.S. pork industry from the Pork Industry Institute from Texas Tech University.

POULTRY SITE
www.thepoultrysite.com
News and articles about poultry.

SAUSAGE MAKING
www.ag.ndsu.edu/pubs/yf/foods/he176w.htm
Information about sausage making from North Dakota State University.

U.N. FOOD AND AGRICULTURE ORGANIZATION ANIMAL WELFARE PORTAL
www.fao.org/ag/againfo/themes/animal-welfare/en/
Access point for a wide range of information related to the welfare of farm animals internationally.

USDA BEEF AND CATTLE INDUSTRY REPORTS
www.ers.usda.gov/news/BSECoverage.htm
Background statistics about U.S. beef and cattle sales, exports, imports, and pricing.

VIRGINIA COOPERATIVE EXTENSION
http://pubs.ext.vt.edu/400/400-803/400-803.html
Information about American, UK, and European beef cattle breeds and biological types.

USDA MEAT AND POULTRY HOTLINE
888-MPHotline (888-674-6854)
mphotline.fsis@usda.gov
Provides answers to food safety questions on subjects such as safe food storage, handling, preparation, product dating, product content, and power outages.

USDA FACT SHEETS
www.fsis.usda.gov/fact_sheets/index.asp
Website providing access to a wide range of food safety and handling fact sheets including meat and poultry preparation, food labeling, and preventing food-borne illnesses.

BOOKS

THE COMPLETE BOOK OF MEAT
Edited by Frank Gerrard and F. J. Mallion, Virtue, 1980.
UK guide to meat production, including markets, agricultural policy, pigs, cattle, sheep, slaughtering, costing, beef and veal cutting, lamb cutting, pork and bacon cutting, poultry and game, offal meats, nutrition, meat storage and safety for meat industry students and professionals.

CULINARY ARTS INSTITUTE KITCHENPRO SERIES: MEAT IDENTIFICATION, FABRICATION, UTILIZATION
Thomas Schneller, Delmar, 2009.
Guide to meat identification, fabrication and uses with illustrated professional meat-cutting techniques for chefs including beef, veal, pork, lamb, and game with foodservice recipes.

CULINARY ARTS INSTITUTE KITCHENPRO SERIES: POULTRY IDENTIFICATION, FABRICATION, UTILIZATION
Thomas Schneller, Delmar, 2010.
Guide to poultry identification, fabrication and uses with illustrated professional poultry-cutting techniques for chefs including chicken, duck, goose, turkey, and game birds with foodservice recipes.

FIELD GUIDE TO MEAT
Aliza Green, Quirk Books, 2005.
Handy pocket-size illustrated guide showing how to identify and choose a large variety of meats including beef, pork, lamb, veal, goat, rabbit, poultry, wild game, sausages, cured meats, and more. Detailed descriptions, selection tips, and color photographs. www.quirkbooks.com/book/field-guide-meat

THE FIFTH QUARTER: AN OFFAL COOKBOOK
Anissa Helou, Absolute Press, 2011.
Offal enthusiast Anissa Helou covers every organ meat from foie gras, blood, marrow, brains, caul fat, testicles, and udder to ears, eyes, tripe, liver, kidneys, cockscombs, giblets, heads, and more.

THE GOOD COOK
Editors of Time-Life Books, 1978, 1979, 1980, 1981, 1982.
This series includes books on beef and veal, lamb, pork, poultry, and variety meats and features information on purchasing, fully illustrated techniques and cooking methods, and a wide variety of recipes reprinted from world cookbooks. Out of print but available from many online used book vendors.

HUMANE LIVESTOCK HANDLING: UNDERSTANDING LIVESTOCK BEHAVIOR AND BUILDING FACILITIES FOR HEALTHIER ANIMALS
Temple Grandin, Storey Publishing, 2008.
This important book covers the natural behavior and temperament of cattle and low-stress methods for moving them about. Also includes detailed humane designs for cattle-handling and slaughterhouse operations.

HUNT, GATHER, COOK
Hank Shaw, Rodale Books, 2011.
Hank Shaw chronicles his passion for hunting wild game and gathering wild foods and shares his experiences in the field and the kitchen, and provides innovative, appetizing ways to prepare wild foods.

THE MEAT BUYER'S GUIDE
NAMP (North American Meat Processors Association), John Wiley & Sons, 2007.
Professional's illustrated, detailed guide to standardized meat cuts covering whole animals, foodservice cuts, and portion cuts for beef, lamb, veal, pork, poultry, game, and processed meat products.

MEAT COOKING ALL'ITALIANA
Savina Roggero, edited and adapted for U.S. readers by Deborah Baker, Thomas Y. Crowell Company, 1975.
Complete guide to Italian meat preparation and cookery with techniques illustrated by line drawings, covering beef and veal, lamb, kid and mutton, pork, poultry (chicken, capon, turkey, duck, goose, guinea fowl, rabbit, and pigeon), game (woodcock, snipe, venison, pheasant, partridge, quail, and hare) with recipes. First published in Italy by Arnoldo Mondadori Editori in 1973 under the title Come Scegliere e Cucinare le Carni (How to Choose and Cook Meat).

THE MEAT WE EAT
John R. Romans, et al., Prentice Hall, 2000.
Standard meat science text and reference book including raising, harvesting, butchering, cutting, processing, and preserving meat.

THE RIVER COTTAGE MEAT BOOK
Hugh Fearnley-Whittingstall, Ten Speed Press, 2007.
The author and chef runs a farm on 60 acres (24.3 hectares) of land in Dorset, England, where he raises a variety of animals that he serves at the River Cottage. This revised American edition includes detailed sections on understanding British meat cuts of beef, lamb, pork, and poultry, including offal, and is based on the author's expertise and respect for nature.

ACKNOWLEDGMENTS

Many people helped me research this book, sharing their experience, colorful stories, useful tips, and valuable advice. I could not have produced this book without the help of James Conboy, owner, and Shawn Padgett, director of purchasing, for the George L. Wells Company, who acted as industry advisors for the book, provided top-quality product for our photo shoots, and patiently answered "just one more question." Working at Wells helped me to gain a better understanding of the demanding and sometimes dangerous work involved in preparing meat. The company has set the standard in the Philadelphia region since its founding over 100 years ago.

I was especially happy to be working with a local company whose owner started out as a meat cutter working the floor. In my many years of dealing with Wells as a chef, my respect grew for their uncompromising efforts to provide the best in meats, poultry, and game trimmed and packaged exquisitely. Special thanks to our expert cutter at Wells, Jerry Booth, a charming, lifelong meat cutter, deservedly respected by his peers for his knowledge, easy-going personality, skill, and ability to teach.

Rochelle Bourgault, the book's stalwart and cheerful editor, kept me on course with her light-hearted humor, clear insights, and careful editing. I give Rochelle extra credit here as she is a vegetarian who was not at all eager to review those large, clear meat-cutting photos. My heartfelt thanks to Clare Pelino, of ProLiterary Agency, for making the successful match between me and Quarry Books. Once again, thanks to Steve Legato, who takes the most beautiful food photographs seasoned with extra helpings of offbeat and sometimes raunchy humor.

Thanks to everyone I interviewed and everyone else I spoke to who helped me understand the incredible complexities of meat production, packing, distribution, aging, cutting, and preparation. Each of these industry leaders shared a different valuable perspective on the issues and choices involved in working with meat from breeders, ranchers, farmers, distributors, butchers, meat cutters, and processors to hunters, importers, restaurateurs, evaluators, chefs, and product creators.

More thanks go to Mariella Giovannucci, co-owner of the incomparable Philadelphia culinary emporium, Fante's, source of many of the special tools used here such as the Italian stainless-steel meat grinder and the Jaccard meat tenderizer. You can't cut good meat without good knives and we had some of the best from Wusthof-Trident of Solingen, Germany. Also, thanks to the Ajinomoto Corporation of Japan for providing samples of Activa transglutaminase, that incredibly useful "meat glue." Thanks to the owners of Senat Halal Poultry for allowing me to visit their busy but clean and efficient halal chicken slaughterhouse in Paterson, New Jersey.

With this book, I hope that enthusiastic home cooks, chefs, and culinarians will gain a nuanced understanding of the world of meat and have the confidence to expand their repertoires to new meats—goat and rabbit, anyone?—and new cuts to help prepare more delicious, more sustainable meats while making the most of their money.

ABOUT THE AUTHOR

Aliza Green, author, consultant, and influential chef, has been fascinated by animal anatomy since ninth-grade biology class where she earned an A+. She is inveterately curious about food ingredients and "meats" fellow food fiends everywhere she travels. Green loves the challenge of working with meat—figuring out which cut to use, how to trim them, and how to cook them in order to maximize flavor, body, pleasing texture, and nutrients. As a voracious reader and self-taught chef, Green learned about working with meat on the job where there were always new meats to cook and skills to polish. *The Butcher's Apprentice* is Green's twelfth book on culinary subjects ranging from beans, fish, and international baking to produce, herbs and spices, and, of course, meat. Green also leads culinary tours with Epicopia, a leader in the field. www.alizagreen.com

ABOUT THE PHOTOGRAPHER

Steve Legato's passion for photography has granted him the humbling opportunity to work with some of the most dedicated, passionate, and creative chefs you've heard of and dozens you haven't heard of—yet. His photography has been featured in *Art Culinaire*, the *New York Times*, *Bon Appetit*, *GQ*, *Wine Spectator*, *Food Arts*, *Travel & Leisure*, and *Wine and Spirits*. He has photographed more than thirty cookbooks, including *!Ceviche!* by Guillermo Pernot, which won a James Beard Award in 2002, and *Nicholas: The Restaurant*, which was nominated for the 2010 IACP Cookbook award for photography.
www.stevelegato.com

INDEX

ALSO AVAILABLE IN THIS SERIES

The Fishmonger's Apprentice
978-1-59253-653-5

The Vintner's Apprentice
978-1-59253-657-3

The Pastry Chef's Apprentice
978-1-59253-711-2

The Brewer's Apprentice
978-1-59253-731-0

ALSO AVAILABLE FROM QUARRY BOOKS

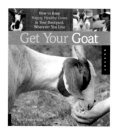

Get Your Goat
978-1-59253-757-0

The Chicken Whisperer's Guide to Keeping Chickens
978-1-59253-728-0

Making Artisan Pasta
978-1-59253-732-7

The Complete Mushroom Hunter
978-1-59253-615-3

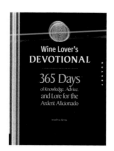

Wine Lover's Devotional
978-1-59253-616-0